THREE SCIENTIFIC REVOLUTIONS

ALSO BY RICHARD H. SCHLAGEL

Animism to Archimedes

Contextual Realism: A Metaphysical Framework for Modern Science

Forging the Methodology that Enlightened Modern Civilization

From Myth to Modern Mind, Vol. I, *Theogony Through Ptolemy*

From Myth to Modern Mind, Vol. II,
Copernicus Through Quantum Mechanics

Seeking the Truth: How Science Has Prevailed over the Supernatural Worldview

The Vanquished Gods: Science, Religion, and the Nature of Belief

THREE SCIENTIFIC REVOLUTIONS

How They Transformed

OUR CONCEPTIONS OF REALITY

RICHARD H. SCHLAGEL

an imprint of Prometheus Books
59 John Glenn Drive, Amherst, New York 14228

Published 2015 by Humanity Books, an imprint of Prometheus Books

Three Scientific Revolutions: How They Transformed Our Conceptions of Reality. Copyright © 2015 by Richard H. Schlagel. All rights reserved. No part of this publication may be reproduced, stored in a retrieval system, or transmitted in any form or by any means, digital, electronic, mechanical, photocopying, recording, or otherwise, or conveyed via the Internet or a website without prior written permission of the publisher, except in the case of brief quotations embodied in critical articles or reviews.

Cover design by Nicole Sommer-Lecht

Inquiries should be addressed to

Humanity Books
59 John Glenn Drive
Amherst, New York 14228
VOICE: 716-691-0133
FAX: 716-691-0137
WWW.PROMETHEUSBOOKS.COM

19 18 17 16 15   5 4 3 2 1

Library of Congress Cataloging-in-Publication Data

Schlagel, Richard H., 1925-
Three scientific revolutions : how they transformed our conceptions of reality / by Richard H. Schlagel.
pages cm
Includes bibliographical references and index.
ISBN 978-1-63388-032-0 (pbk.) — ISBN 978-1-63388-033-7 (e-book)
1. Science—History. 2. Science—Greece–History. 3. Science—History—20th century. 4. Science—Philosophy. I. Title.

Q125.S4145 2015
509–dc23

2015001969

Printed in the United States of America

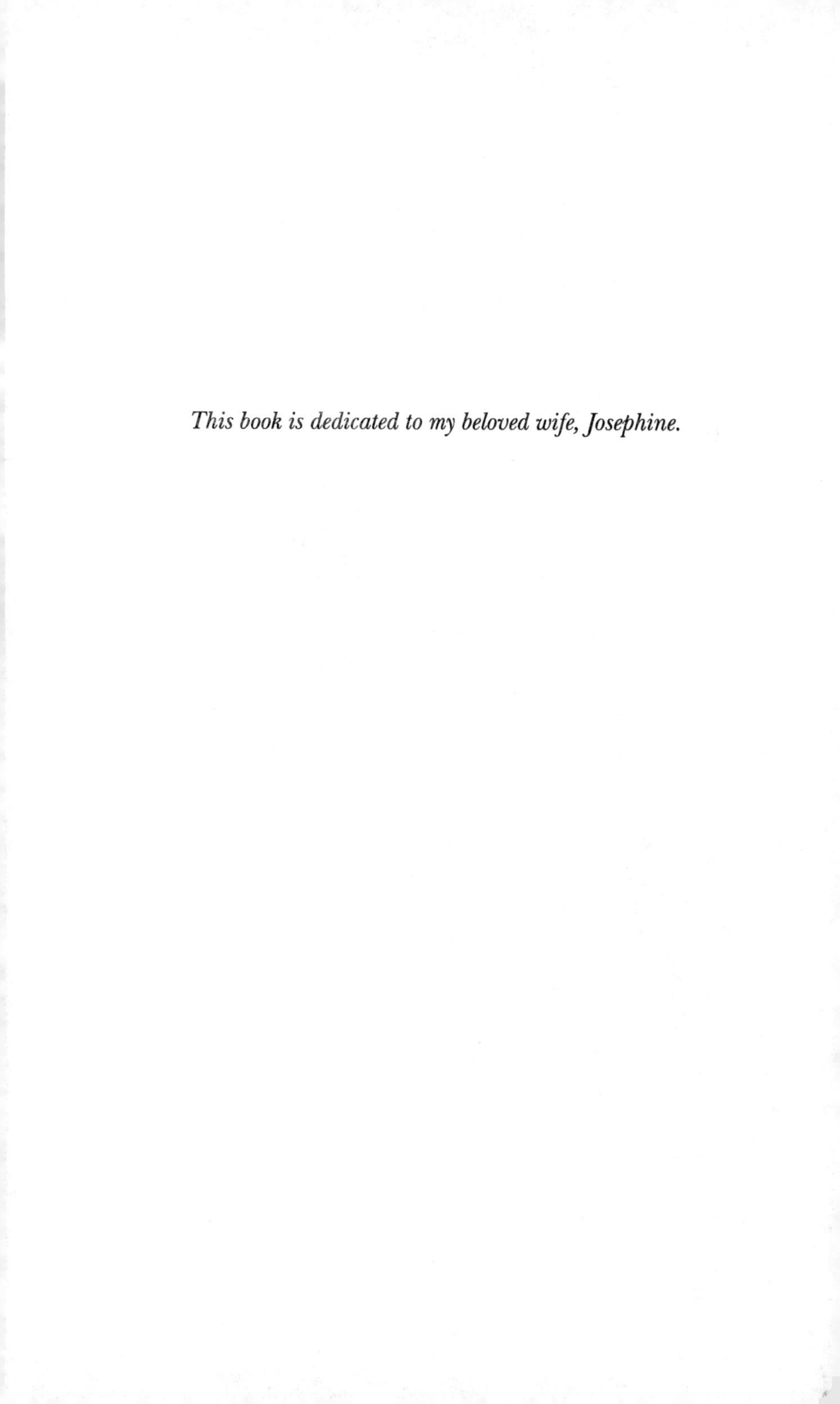

This book is dedicated to my beloved wife, Josephine.

CONTENTS

PREFACE

I am dismayed that there is such a large number of Americans who are so ignorant about or indifferent to the contributions of science to our understanding of the universe and human existence that they continue to believe incredible religious doctrines. As examples, all seven of the republican candidates who ran in the 2012 presidential primary declared their *dis*belief in evolution on the grounds that there was insufficient evidence, even though probably no scientific theory has more supporting empirical confirmation considering the discoveries of transitional fossil remains extending back millions of years and the ancestral genealogical record replicated in the human genome. What other *explanatory* religious evidence can Christians offer to explain the transitions and diversity of species?

Moreover, not one of the basic theological doctrines or rituals of Christianity any longer has any rational credibility: for example, Mary's reputed immaculate conception in terms of what we now know about bisexual reproduction, along with specific references in several Gospels to Jesus' brothers and sisters; or Jesus' virgin birth, which would have lacked the essential twenty-four male chromosomes; his miracles such as walking on water, multiplying loaves of bread, raising Lazarus from the dead; and the discrepant Gospel accounts of his reappearing after his crucifixion in human form that supposedly confirmed his divinity. One of the essential rituals of Christianity is the Eucharist or Holy Communion during which the bread and wine, after being consecrated, are supposedly transubstantiated and consumed as the flesh and blood of Jesus, which is an impossible chemical transmutation (the scientific term) of substances, though still literally affirmed by the Catholic Church despite being denied even by noted Catholic theologian John Wycliffe as early as the fourteenth century.

Hoping to mitigate this ignorance, I describe how major scientific advances produced three past revolutionary transformations in our conceptions of physical reality and human existence, beginning with the ancient Greeks and progressing to the present, refuting these implausible beliefs. In the final chapter I offer a summary of recent discoveries and forecasts of future advances as prologue to the fourth transition.

The first transformation was the Hellenic and Hellenistic replacement of the previous mythical and theological attempts to explain the origin and nature of the universe and human existence with a partial empirical-rationalistic method of inquiry that was the precursor of modern classical science. Aristotle's philosophy was still prevalent as late as the seventeenth century as indicated in Isaac Newton's assertion that "Aristotle and Descartes are my main adversaries," while each of the founders of modern classical science cited ancient Greeks as their predecessors.

The second revolution was initiated by Copernicus's rejection of the venerable geocentric conception of the universe stating in the introduction to his *De revolutionibus orbium coelestium* (On the Revolutions of Heavenly Spheres), published in 1543 and addressed to the ecclesiastic authorities, that even though it may seem absurd, since others had been granted the right, he too would assign a circular motion to the earth around the sun indicating that he could "correlate all the movements of the other planets . . . with the mobility of the earth." It was this work that motivated Johannes Kepler to pursue astronomical research leading to his discovery of Tycho Brahe's observations of the elliptical orbit of Mars, which convinced him that it was not the earth's centrality but the sun's emanations that produced the elliptical motion of the planets that enabled him to formulate his three astronomical laws accurately describing their orbital velocities and distances from the sun. This culminated in his conception of a "clockwork universe" with "gravity" as the key astronomical force producing their motions. Both conceptions were introduced prior to Newton who is usually given credit for formulating them!

These discoveries coincided with Galileo's amazing telescopic observations disclosing the incredible resemblance of the moon's presumed astral surface to the earth's terrestrial geology and the discovery of the "rings of Saturn" circling Jupiter, the "phases of Venus," along with the seven stars known as the "Pleiades." In addition, he observed innumerably more fixed stars implying that the universe was vastly more extensive than previously thought. His dramatic drawings of these observations showing the similarity of the moon's celestial surface to the earth's terrestrial terrain in the *Sidereus Nuncius* (The Starry Messenger) fueled the controversy as to whether the earth or the sun was in the center of the solar system.

It also was in the *Sidereus Nuncius*, after analyzing the difference between the optical nature of astronomical observations and ordinary perceptual sensations such as colors, sounds, and heat, and influenced by his improved microscopic observations (analogous to the impact of his telescopic discoveries) that Galileo introduced the crucial causes later attributed to waves or "insensible particles." As he states in *The Assayer*: "[m]any sensations which are deemed to be qualities residing in external objects have no real existence except in ourselves. . . . "*

Though defined in terms of observable *properties* such as sizes, shapes, masses, and motion, the insensible particles were devoid of sensory *qualities* such as colors, sounds, smells, tastes, or heat that are produced in us by their impact on our senses. This distinction first proposed by the ancient Greek philosophers, Anaxagoras, Leucippus, and Democritus and known as imperceptibles or "atoms," had been eclipsed owing to Empedocles' acceptance of the four elements of fire, earth, air, and water, as being indestructible and thus basic. Galileo's renewed distinction proved one of the most controversial conceptions in modern science and philosophy during the nineteenth and twentieth centuries.

Given the disputes over his astronomical observations and whether the earth or the sun was the center of the solar system, Galileo decided to go to Rome to gain permission from Pope Urban VIII to publish a book on the evidence for the two worldviews. Urban gave his consent provided that Galileo treated the geocentric and the heliocentric positions impartially; but once the *Dialogue Concerning the Two Chief World Systems* (its English title) was published, obviously revealing Galileo's preference for the heliocentric system, the pope concluded that he had been disobeyed.

Thus Galileo was brought before the Inquisition of the Holy Office, was tried for heresy and found guilty, forcing him on his knees before the Commission to recant his support for heliocentrism on threat of imprisonment, torture, and perhaps death (in 1600 at the time that Kepler and Galileo were presenting their major discoveries Giordano Bruno was ordered burned at the stake in Rome by the Catholic Church for advocating such heretical ideas as an infinite universe). Yet insisting it was not his intention to support heliocentrism, Galileo was never punished but put under house arrest in his country villa in Arcetri for the rest of his life and ordered not to discuss astronomy or receive any visitors under penalty of imprisonment.

It was there that he wrote his second most famous work entitled in English, *Dialogues Concerning Two New Sciences,* in which he described his incline plane experiments demonstrating that a ball rolling down a smooth incline plane in equal times traverses distances proportional to the odd numbers beginning with 1:1, 3, 5, 7, 9, etc., so that the square roots of the successive sums of the odd numbers give the progressive *times* of the descent. Dying a few years after his final publication, he was interred in the splendid church of Santa Croce in Florence and though the Grand Duke originally forbad any ornamental additions to his tomb, now there can be seen an exquisite sepulcher comparable to the one of Michelangelo across from his.

These discoveries culminated in Newton's deterministic cosmological theory of absolute space and time, along with his universal law of gravitation applying to both the celestial and terrestrial worlds, thus refuting their historic qualitative distinction. Moreover, his corpuscular-mechanistic universe incorporating Galileo's submicroscopic particles set the agenda for much of the research during the following eighteenth century, known as the "Age of Enlightenment," and the nineteenth and twentieth centuries.

Among his other discoveries, Newton demonstrated with his prismatic experiments that light is not homogeneous, because when refracted through a prism it disperses into a spectrum of specific colors that he describes as rays radiated as corpuscles, contrary to the prevailing wave theory of light. He also proved the finite velocity of light. Yet as innovative as they were, these modern classical physicists still believed in a creator God, but (like Albert Einstein) they claimed that once the universe was created God did not intervene in it and so these physicists were usually known as deists rather than theists.

The third revolution began in the latter nineteenth century with the discovery of such subatomic particles as the electron, proton, neutron, and proof of electromagnetism, along with Charles Darwin's later evolutionary theory of natural selection. The twentieth century brought about further astonishing developments in particle physics culminating in two fundamental classes of particles, the fermions and bosons, plus the discovery of strong and weak nuclear forces. It also encompassed innovative advances in geology, paleontology, chemistry, biology, medicine, neurophysiology, and genetics, along with creating such significant theories as Max Planck's quantum mechanics, Einstein's theories of relativity, Niels Bohr's solar model of the atom, Werner Heisenberg's introduction of matrix mechanics and the uncertainty principle.

It also comprised the vastly improved telescopes such as Edwin Hubble's telescope at Mount Wilson's Observatory in Los Angeles and even more powerful telescopes that by the end of the century disclosed a much more extensive universe with about 125 billion galaxies each consisting of trillions of stars, as well as the dramatic lunar landing confirming the geographical similarity of another planet to the earth. Furthermore, there was evidence of the accelerating expansion of the universe following the big bang and the existence of black holes and dark energy about which we know very little and yet they apparently compose around 95 percent of the matter of the universe. Also, since the big bang itself presumably had a cause, this suggests that it must be an offshoot of another universe giving rise to the concept of multiverses.

The final chapter discusses the current scientific advances that are a prologue to the impending fourth revolutionary transformation in our conception of reality and way of life. These include the vastly extended telescopic observations owing to the creation of such recent powerful telescopes as the James Webb Space Telescope and the Kepler Space Telescope, along with the Search for Extraterrestrial Intelligence (SETI) to search among the trillions of exoplanets for evidence of the environmental conditions that would allow for the existence of living creatures and especially extraterrestrial intelligence. They also comprise interplanetary space explorations to seek a new home for terrestrial human beings when our planet becomes either uninhabitable or consumed by the eventual increase in radiant heat from the sun due to the latter's future atmospheric changes.

In addition, there is the fantastic progress in computer science and robotics that researchers predict will permit human beings to have each of their cranial neurons and synaptic connections replaced by electronic components and then encoded in a computer program or installed in the head of a robot and thus achieve a weird kind of programmed existence or robotic longevity. Equally fantastic are the anticipated medical advances plus genetic discoveries that will eliminate most health problems, along with the detection of the "aging genes" that will greatly extend our lifespan if there can be found another habitable planet to travel to and live on.

The chapter also describes the latest attempt at a final unified theory, namely, string theory, which has not produced any empirical evidence and thus is unlikely to be a concluding theory thereby raising the question as to whether it is even possible. Given the seemingly unlimited dimensions and diversity of the universe, the previous scientific assumption that a final unified theory (that eluded Einstein) is attainable now seems to me to be improbable. If, due to limited knowledge and funds, no such theory is feasible, this obviously does not detract from the fact that Christianity and other world religions, along with philosophical metaphysical systems, are no longer credible in contrast to the formulation and adoption of the scientific methodology that produced the described transformations in our worldviews. In its place I have suggested the theory of "Contextual Realism" in a previous book by that title and mentioned later in the present book.

Furthermore, I find more encouraging and promising the advances in genetics, such as heterochromatin, that will enable scientists to identify the essential genes at least partially responsible for the terrible cruelties, iniquities, and suffering pervading human existence. Just as geneticists have discovered the particular genes or systems of genes and the cellular fluids that direct the causes of human physical illnesses, they are now close to detecting the genetic causes that produce the greatest human atrocities: murders, rapes, assaults, repressions, injustices, deceptions, etc. Detected and deactivated they would enable us to improve human nature thereby supplementing or even replacing religion, parental guidance, and social laws as the major determinate influences directing human behavior. Considering the contentious, devastating, and depressing past history and present state of the world situation, I find the prospect of genetically improving human nature the most promising and admirable hope for the future, vastly preferable to being electronically programmed in a computer or installed in a robot.

Chapter I

THE FIRST TRANSITION OWING TO THE NATURAL PHILOSOPHIC INQUIRIES DURING THE GREEK HELLENIC AND HELLENISTIC PERIOD

Considering that the United States emerged as the dominant world power after World War II due to its superior armaments, which were based on its advanced scientific and technological developments, and also to its being the freest and most prosperous country after defeating Russia in the Cold War, it is appalling how little most Americans know about and appreciate the reasons for these achievements—that it was the ancient Greeks who first initiated the scientific method of inquiry that contributed so greatly to America's ascendance while the conception and adoption of democracy also was first introduced in Athens by Cleisthenes in 508 BCE. According to Robin Lane Fox, an ancient historian, in his *The Classical World*,

> in the spring of 508 BC . . . Cleisthenes proposed . . . that the [Athenian] constitution should be changed and that, in all things, the sovereign power should rest with the entire adult male citizenry. It was a spectacular moment, the first known proposal of democracy, the lasting example of the Athenians to the world.[1]

As supporting evidence of these two crucial influences, science and democracy, astrophysicist Carl Sagan stated in his incredibly informed book *The Demon-Haunted World: Science as a Candle in the Dark*: "At the Constitutional Convention of 1789 John Adams repeatedly appealed to the analogy of mechanical balance in machines . . ."; "James Madison used chemical and biological metaphors in *The Federalist Papers*"; and Thomas Jefferson, who described himself as a scientist, wrote in the Declaration of Independence, "that we all must have the same opportunities, the same 'unalienable' Rights,"[2] though sadly this did not include women and slaves. As Jefferson adds:

> In every country, we should be teaching our children the scientific method and the reasons for a Bill of Rights. With it comes a certain decency, humility and community spirit. In the demon-haunted world that we inhabit by virtue of being human, this may be all that stands between us and the enveloping darkness. (p. 434)

In this book I shall describe the three past revolutionary scientific transitions that radically transformed our conceptions of the universe and human existence. I also argue that given the enormity and complexity of the universe the traditional scientific goal of a "unified final theory" should be replaced by the theoretical framework of "contextual realism."[3] Rather than seeking a *final* theoretical framework to explain all empirical evidence as most scientists of the past intended, we should realize that such inquiries are conducted within successively deeper and expanding *conditional* but nonetheless *real* physical contexts of the universe that appear to be endless.

Turning to the first scientific transformation of our conception of reality, while the Egyptians and Mesopotamians had made significant contributions in astronomy, mathematics, biology, and medicine that antedated the scientific inquiries of the ancient Greeks, it is generally conceded that it was the latter who first began a *systematic* attempt to attain a more empirical-rational understanding of the universe by replacing the previous mythological and theological accounts with empirical observations, logical and mathematical reasoning, and rational explanations.

For instance, it was the Greek Milesians Thales, Anaximander, and Anaximines who, in the sixth century BCE, rejected a divine creator of the universe for naturalistic explanations in terms of Water (Thales), an Unbounded (Anaximander), and an Air-Substrate (Anaximines) and adopted such ordinary explanatory principles as "separating off" or "condensation and evaporation" to explain how our current universe came to be from that original state. Though an admirable effort, this attempted *unified explanation* is now referred to as the "Ionian fallacy."

Another extremely gifted person whose influence extended throughout the centuries (string theory in physics is a modern example) was the Ionian philosopher Pythagoras of Samos, also from the sixth century, who was a musician, mathematician, astronomer, mystic, and founder of the Pythagorean philosophical and religious school in Croton. Reputed to be an accomplished lutenist, this facilitated several of his unique mathematical discoveries, the first being that the intervals of musical scales in which the consonances and successive octaves could be expressed in numerical ratios comprising the first four integers. This was followed by his speculation that the motion of the planets emits a musical harmony called the "Music of the Spheres," though too remote to be heard by human ears.

Among his other mathematical discoveries were irrational numbers, the Pythagorean theorem, the tetractys (a triangular figure of four rows of numbers that add up to the perfect number ten), and that spatial configurations can be created from "arithmogeometric units"—e.g., an extended line drawn from two points, plane figures such as triangles and rectangles from several lines, a circle from a joined curved line, and three-dimensional spatial objects such as pyramids cubes, spheres, and complex polyhedra from plane figures. As Aristotle states, based on these inquires "the Pythagoreans . . . construct the whole universe out of numbers—only not numbers consisting of abstract units: they suppose the units to have spatial magnitude."[4]

Thus the Pythagoreans were able to represent the four elements of the physical world—earth, air, fire, and water—by four polyhedra: the earth by the 4-sided pyramid or tetrahedron, air by the 6-sided cube, fire by the 8-sided octahedron, water by the 20-sided icosahedron, and the universe itself by the 12-sided dodecahedron. Because Plato apparently assigned different polyhedra to the four elements, explaining their disintegration and reconfiguration as due to the separation and recombination of their constituent plane figures, they came to be known as "the five Platonic solids." Kepler in the early seventeenth century began his astronomical theorizing in his *Mysterium Cosmographicum* (The Cosmographic Mystery) with the five polyhedra of Pythagoras perhaps as revised by Plato. Other of their astronomical contributions also were extremely important, such as Eudoxus of Cnidus who made the determination of the solar year to be 365 days and five hours, along with originating the long-prevailing view that the celestial bodies revolve on a series of concentric spheres with the earth in the center.

His pupil Callippus of Cyzicus increased his number of spheres to thirty-four to account for certain astronomical irregularities that were adopted by Aristotle. But Philolaus of Croton, in 259 BCE, astutely assigned "an oblique circular motion" to the earth around a central fire while Heraclides of Pontus and Ecphantos of Syracuse attributed to it an axial rotation from west to east to explain the apparent rising and setting of the sun, along with determining that Mercury and Venus revolve around the sun. This culminated in Aristarchus of Samos's prescient sun-centered astronomical theory in the third century BCE, though eclipsed by Ptolemy's geocentrism until Copernicus's adoption of heliocentrism.

These celestial innovations were complemented by such empirical theories as Empedocles' conception of the four elements, earth, air, fire, and water, as basic; Anaxagoras' rejection of Empedocles' four elements as too limited, declaring that the original mixture consisted of an infinite number of infinitely divisible particles that were representative of all the diversity of things, but too minute to be discernable except for air and aither; Leucippus' and Democritus' astute atomic theory that the underlying matter of the universe consisted of solid, indivisible, insensible particles that varied in their size, shapes, solidity, and motions, excluding sensory qualities.[5]

However, deriding such empirical explanations Plato, in his famous "allegory of the cave," described sensory knowledge as mere reflections of the imperfect material objects in the physical world or "Receptacle," declaring that mathematics could free one from these perceptual illusions to ascend to the intelligible world of perfect archetypes, the "Realm of Forms," culminating in the "Form of the Good" and the "Demiurge." Apparently the latter was the creator of the real world by imposing the ideal archetypes on the imperfect Receptacle.[6] It was Plato's philosophy that was the most influential during the medieval period because of its easy conformity with Christianity, interpreting his Demiurge as God.

Yet it was not Plato's philosophy but that of his pupil Aristotle that would prove the most dominant from the thirteenth to the seventeenth century following the syntheses of his philosophy with Christianity by Thomas Aquinas. Rejecting Plato's methodology that relied on mathematics for attaining knowledge of the Forms because Aristotle thought it only applied to abstract magnitudes, not to the empirical world, he created the formalism of logic for deducing specific physical properties from empirical premises stating their *genus* and *species* derived from empirical inductions.

There were three major factors explaining the greater acceptance of his philosophy. First, that his basis of knowledge relying on ordinary perceptions, as interpreted within his schema of the four causes, made it less abstract and idealistic and more empirically amenable. These included the "material cause" (the physical composition of objects eventuating in "prime matter)"; the "formal cause" delineating the "species, genus, and definitions to which it belonged"; the "efficient cause" that produces the interactions and changes in nature; and the "final cause" or "end of which" an object or process aims. This final cause involving the actualization of an inherent potentiality added to the appeal because it suited the general conception at the time that all events had an innate purpose.

It was in *The Prior Analytics* that Aristotle created syllogistic logic as his methodology for proving the existence of specific physical properties and efficient and final causes by *deducing* them from *inductive general premises* specifying the particular genus, species, or definition of the object. As illustrated in his classic examples: one can prove that Socrates is mortal in the syllogism "All men are mortal, Socrates is a man, therefore Socrates is mortal" or demonstrating why, in contrast to the stars, the planets do not twinkle, from the premise "No proximate celestial body twinkles, the planets are such proximate bodies, therefore the planets do not twinkle." He concluded that since the middle terms, such as 'men' and 'proximate celestial body' conjoining the premises provided the proof, they were not merely verbal connections but the *actual causes* of the conclusion stating that "in all our inquiries we are asking either whether there is a 'middle' or what the 'middle' is: for the 'middle' here is precisely the cause, and it is the cause that we seek in our inquiries."[7]

Yet it is not just this formal methodology that accounted for Aristotle's tremendous influence, but also the extraordinary range of his research covering nearly every known area of human experience at the time. This includes, in addition to his writings on ethics, politics, rhetoric, poetics, categories, and logic, works "On the Heavens," "On the Soul," "Metaphysics," "Physics," "Generation and Corruption," "Memory, Dreams, and Prophesying," along with the "History, Parts, and Generation of Animals." I think it can be said that no other thinker ever matched Aristotle in the range and quality (for the time) of his extensive research. Charles Darwin was so impressed by his biological writings that he wrote: "Linnaeus and Cuvier have been my two gods [. . .] but they were mere school-boys compared to old Aristotle."[8]

The third factor responsible for his immense influence was his geocentric cosmology that seemed most congruent with our ordinary observations with its distinction between the perfect celestial and imperfect terrestrial worlds involving their contrasting natures and motions: the celestial or heavenly bodies consisting of an aetherial substance and having inherent circular and uniform motions while the terrestrial world consisted of the four Empedoclean elements (earth, air, fire, and water), each with its inherent rectilinear motion upward or downward on the stationary earth. Thus it was Aristotle's more common-sense cosmological system, as emended by Ptolemy and defended by the Scholastics, that generally prevailed from about the thirteenth to the seventeenth centuries and was mainly the system that had to be replaced by the scientific inquiries of Nicholas Copernicus, Johannes Kepler, Galileo, Christiaan Huygens, Robert Boyle, and Isaac Newton, the latter declaring that his "two main adversaries were Aristotle and Descartes."

Before turning to the next historical period and major scientific contributions in ancient Greece, some mention should be made of the secular philosophy and influence of Epicurus (341–270 BCE) as poetically transposed and popularized by the Roman poet Lucretius (ca. 96–ca. 55) owing to most of Epicurus' works being destroyed by the burning of the library in Alexandria.

Born on the Island of Samos, Epicurus left for Athens to study the philosophies of Democritus and Plato and where he later purchased a house and garden that did not serve like the more prestigious academic institutions of Plato's Academy or Aristotle's Lyceum, but as a sheltered enclave where his followers, including women and slaves, could listen to his inspired teachings and discuss his numerous books. Primarily concerned with the ethical question of how to live a tranquil life in a world of conflicts, adversity, and suffering, he accepting that the gods existed due to the universal belief in them and the images (*eidola*) they conveyed in dreams and mystical experiences, yet he denied they exerted *any influence* on human affairs, being divine and involved in their own peaceful existence.

He believed that the universe, including our bodies and our souls, consists of atoms and the void. He even introduced the prescient theory that the various sizes and shapes of the atoms could be explained by their being composed of "internal minima," which, like the present-day quarks, help account for their physical characteristics yet cannot exist separately or independently. Though adopting Democritus' atomism he denied his strict determinism, introducing a spontaneous "swerve" in the formation of the world to account for its diversity and novelty and to explain free will. Denying religions as superstitions, he believed in an infinite, eternal universe that did not require a creator. Since souls consist of atoms they do not outlive the body and thus one does not have to fear any retribution after death which is the termination of life.

Although Epicurus' ethics was based on the fact that human beings are primarily driven by their desire for pleasure and avoidance of pain, he was aware that not all pleasures are desirable, many are accompanied by painful consequences, thus they must be chosen wisely. The following verse (as translated from the Greek) etched on a wall in Herculeum expresses his ethical philosophy.

There is nothing to fear in God
There is nothing to feel in death;
What is good is easily procured
What is bad is easily endured.

It is thanks to the recovery by Poggio Bracciolini of the epic poem of Lucretius, *De rerum natura* (On the Nature of Things) in a remote monastery library in Herculaneum in 1417, that we have some glimpses into Epicurus' philosophy as presented in Lucretius's extraordinary rendition. Although written in eloquent hexameter verse rather than philosophic prose, it represents the most advanced, rational, and realistic worldview of ancient philosophy. As Fox states in his *The Classical World*, previously cited:

> I have tried in this book to tell a little known but exemplary Renaissance story, the story of Poggio Bracciolini's recovery of *On the Nature of Things*. The recovery has the virtue of being true to the term that we use to gesture toward the cultural shift at the origins of modern life and thought: a renaissance, a rebirth, of antiquity. One poem by itself was certainly not responsible for an entire intellectual, oral, and social transformation—no single work was, let alone one that for centuries could not without danger be spoken about freely in public. But his particular ancient book suddenly returning to view made a difference. (p. 11)

Over seventeen centuries would elapse before it was confirmed that the ordinary world was actually composed of what Epicurus still referred to as Democratean atoms but Lucretius called "first things" or "the seeds of things." They were eternal, unchanging, imperceptible, ultimate particles that exist in an infinite spatial void whose constant motions and interactions create the great diversities of nature along with their destructions, since everything thus created is perishable except the particles themselves. As this includes our souls along with our bodies they, too, decompose when we die, even though they are of a finer nature, and thus there is no afterlife. In this way Epicurus and Lucretius eliminated the suffering and retributions that were the greatest fears instilled by religions, especially the Christian Inquisition, one of the most terrifying periods in history. And there is no purpose to existence, just the natural occurrences produced by the imperishable particles.

Though he was not an atheist since, like Epicurus, Lucretius did not deny the existence of the gods, claiming that they were entirely too exalted and involved in their own affairs to be concerned with humankind. Rejecting an ethics based on divine moral principles and reinforced by the threat of eternal damnation or the reward of a beatific afterlife that they regarded as delusional, they defined the highest ethical principle as the "enhancement of pleasure and reduction of pain," according to Fox (p. 195).

Though aware that the needs and desires of mankind, such as the satisfaction of sexual drives, nutritional needs, and shelter, along with the gratifications that come with the attainment of prestige, power, and fame must be fulfilled to some extent Lucretius, like Epicurus, stressed moderation. They both considered that a tranquil existence with reasonable and wholesome pleasures is the ultimate goal in life and more easily attained than a voluptuous, hedonistic, despotic, God-fearing life. Having declared that the original particles move in a random, deterministic way but realizing that ethics requires free will Lucretius, as did Epicurus, introduced a "swerve" to allow some novelty in nature and freedom of the will. Fox again presents a very concise but accurate summary of Lucretius's poetic rendition of the Epicurean philosophy.

> The realization that the universe consists of atoms and void and nothing else, that the world was not made for us by a providential creator, that we are not the center of the universe, that our emotional lives are no more distinct than our physical lives from

> those of all other creatures, that our souls are as material and as mortal as our bodies—all these things are not the cause for despair. On the contrary, grasping the way things really are is the crucial step toward the possibility of happiness. Human insignificance—the fact that it is not all about us and our fate—is, Lucretius insisted, the good news. (p. 199)

What an enlightened conception of reality that still is not accepted by a majority of Americans and other civilizations, though increasingly acknowledged by Europeans.

Returning to our historical narrative, Aristotle's death in 322 BCE followed by that of Alexander the Great a year later, coincided with the usual date 323 BCE given for the termination of the Hellenic classical period. This was succeeded by the Hellenistic Age, conventionally dating from the accession of Alexander the Great to the Macedonian throne in 336 BCE after the death of his father King Phillip, to the death of the Egyptian queen, Cleopatra VII, in 30 BCE, although including some later Hellenistic scholars. The year before his untimely death Alexander had designated the port city Alexandria in Egypt as his namesake, but his empire having been divided into three portions after his death, with General Ptolemy taking control of Alexandria, the growth of the port into the largest and most prestigious Hellenistic city is due to the wise governing of the Ptolemaic dynasty that endured for about three centuries.

It was the Ptolemies who created its famous Museum that became the center of research in astronomy, mathematics, physics, engineering, anatomy, and medicine that eventually eclipsed Plato's Academy and Aristotle's Lyceum as the world's center of learning. Adjacent to the Museum was the equally famous Royal Alexandrian Library that, according to the earliest account, was built during the reign of Ptolemy I Sorter (ca. 367–ca. 283 BCE), but organized by a prestigious student of Aristotle, Demetrius of Phaleron, that became the greatest library of the ancient world.

This period has been referred to as the "first great age of science," surpassing the achievements of the Hellenic Greeks because their scientific investigations and discoveries were less speculative, conforming more to and thus the forebear of, modern classical science.[9] Unlike Aristotle whose scientific works have all been disproved despite their profound historical influence, some of the scientific and mathematical contributions of the Hellenistic thinkers are still valid. Listed in their chronological order they include Euclid, the most famous of the Alexandrian mathematicians who wrote in the third century BCE. Though he began his mathematical studies in Plato's Academy, he wrote his famous *Elements of Geometry* while in Alexandria, acclaimed as the most widely read book in history except for the Bible and extolled by the young Einstein as the model for scientific reasoning.

The second most renowned mathematician, also of the third century BCE, was Archimedes who lived in Syracuse, Sicily, but visited Alexandria two years after the death of Euclid. His outstanding contributions include his method of exhaustion anticipating differential calculus, discovery of the law of specific gravity, and formulation of the principles underlying many technological inventions such as the lever, the pulley, and the tubular screw used to pump water from wells and mines. It is reputed that his pulleys were so powerful that during the siege of Syracuse he was able to attach them to the bows of Roman ships lifting and twisting them out of the sea casting the terrified crew overboard. A third famous Hellenistic mathematician was Hipparchus of Nicaea, who lived in the second century BCE and is especially known for founding plane and spherical trigonometry.

We previously discussed the contributions of Aristarchus of Samos, who wrote in the third century BCE and is referred to as the "Hellenistic Copernicus." Yet the only evidence we have of this is in Archimedes' description in the "The Sand-Reckoner."

> Now you [Kind Gelon] are aware that "universe" is the name given by most astronomers to the sphere whose centre is the centre of the earth. . . . But Aristarchus of Samos brought out a book consisting of some hypotheses. . . . [such as] that the fixed stars and the sun remain unmoved, that the earth revolves about the sun in the circumference of a circle, the sun lying in the middle of the orbit.[10] (Brackets added)

Although partially preceded by the Pythagoreans and Philolaus, the quote's introduction of the heliocentric worldview by Aristarchus is one of the most striking in the history of astronomy.

The other significant astronomer previously referred to is Hipparchus whose major works were written in Alexandria. His varied contributions included the design of the astrolabe, an authoritative star chart, recognition of the precession of the equinoxes, and very exact measurements of the moon's diameter and distance from the earth. However, as important as these previous contributions were, it was Ptolemy's *Almagest* (the title of his major work later given by the Arabs) written in Alexandria in the second century CE whose astronomical system with its epicycles, eccentrics, and equants, to accommodate the astronomical observations that did not fit the spherical orbits and uniform motion of Eudoxus' celestial system, that eclipsed Aristarchus' heliocentrism and that prevailed until its rejection by Copernicus.

But, it was Eratosthenes, the famous librarian of Alexandria in the third century BCE, having acquired a notable reputation as an astronomer, mathematician, geographer, and philologist, who rivaled Aristotle as "the most learned man of antiquity." Known for his invention of the "Sieve of Eratosthenes" for deriving prime numbers and his astute geometrical studies, his mathematical gifts facilitated his remarkable geographical discoveries. He drew the most accurate map of the world for the time showing the circumference of the earth divided into latitudes and longitudes, proposed that the oceans were so united that it would be possible to reach India by sailing west and, most importantly, introduced an ingenious mathematical method for measuring the circumference of the earth within an accuracy of 200 miles.

Another amazing researcher in Alexandria of the third century BCE was Herophilus of Chalcedon, one of the first to practice human dissection and consequently is considered the outstanding anatomist of ancient Greece. Among his discoveries was that the arteries carried blood from the heart to all parts of the body, the usefulness of the pulse in diagnosing various illnesses, and that by dissecting the brain various bodily functions could be correlated with specific brain regions, a remarkable discovery for the time. He was succeeded by Erasistratus, who is said to have practiced vivisection in Alexandria also in the third century. He, too, is famous for having discovered the functions of various body parts, such as the valves in the heart and distinguishing between the arteries and the veins detecting their interconnection. (For more on ancient medicine see volume one of my book *From Myth to Modern Mind: A Study of the Origins and Growth of Scientific Thought.*)

Still, despite the belief that revealed scripture was far superior to scientific knowledge, the best known of the physiologists is Galen of Pergamum, who in his youth studied in Alexandria and other centers of learning acquiring the vast knowledge of medicine for which he is famous and then settling in Rome for the rest of his life in the second century CE. Three reasons account for his prominence: (1) his encyclopedic knowledge; (2) that while most of the works of the previous Ionian scholars were lost in the various fires that destroyed the Royal Library in Alexandria his, fortunately, were preserved; and (3) that his own physiological research was so advanced that it prevailed until replaced by the work of Andreas Vesalius in the sixteenth century.

His major achievement was his description of the physiological organs and functions integrating the circulatory, respiratory, and nutritive systems. He described how the "cosmic pneuma" was inhaled through the trachea, carried to the lungs, and then transmitted by the *Vena arterials* to the left cavity of the heart where it was mixed with the blood. The blood itself was derived from nutriment taken by the portal vessel from the intestines to the liver where it was converted to venous blood by combining with a second spirit or pneuma, called "natural spirit," which is viewed as essential for life.

The combination of natural spirit and nutriment composing the venous blood in the liver is then dispersed by the veins throughout the venous system. When some of this venous blood is carried to the right cavity of the heart it divides into two portions, the larger discharging its impurities back into the *Vena arterials* where it is carried to the lungs and breathed out with the remaining purified portion returning to the venous system. This smaller portion slowly flows through tiny vessels passing the septum dividing the two sides of the heart entering drop by drop into the left side. There they mix with the outside pneuma or natural spirit entering the trachea producing a third, higher type of pneuma, the "vital spirit": the terms 'vitalism,' 'vital principle,' and Bergson's *élan vitale* also were used in the late nineteenth and early twentieth centuries to counter the claim that strictly mechanistic or physiological theories could fully explain evolution. The dark venous blood is then transformed into a bright arterial blood and dispersed via the arteries throughout the body and to the base of the brain where it is activated by another pneuma, an "animal spirit," animating the body.

This extraordinary explanation was facilitated by his dissection of the Barbary apes whose anatomy closely resembles humans. It also illustrates the remarkable progress made in devising better investigative methods and more accurate scientific explanations since Aristotle (who located consciousness in the heart). An ardent teleologist, Galen believed that everything was ordained by God, which motivated his research and is one reason his system was so popular during the Middle Ages.

Brief mention should be made of three other contributors to Alexandrian research. First is Hero or Heron of Alexandria who lived in the first century CE and is known for his ingenious technological inventions in pneumatics and mechanics. These include a globe with attached jets through which the steam from an underlying boiling caldron in successively passing through the jets causes their rotation, a precursor of the steam engine; a cogwheel turned by a twisted screw; multiple pulleys; and a Dioptra for measuring the angles and heights of distant objects.

Second is Rufus of Ephesus, who also lived in the first century CE and made crucial advances in understanding the structure and functioning of the eyes, some of his nomenclature still used today. Third is Diophantus, who lived in the second century CE and is recognized for his contributions to algebra and for introducing signs for minus, equality, unknowns, and powers used to solve various algebraic functions. While these were important discoveries, progress in algebra remained far behind the advances in geometry made by Euclid and those in trigonometry made by Hipparchus of Nicaea.

Had these inquiries continued, modern classical science would not have had to wait nearly two millennia before its resumption. Despite the Romans' extraordinary gifts for engineering and architecture as seen in their splendid aqueducts, temples, baths, and colosseums; for creating some of the world's greatest literature in the writings of Cicero, Virgil, Horace, Ovid, and Pliny; for their remarkable artistic talents displayed in the lovely frescoes in Livia's villa and recovered in Pompeii and Etruria; and for their interest in reading resulting in their creating beautifully designed public libraries throughout the empire, one cannot cite a single outstanding mathematician or natural philosopher who was not Greco-Roman.

After the rise of Christianity and the transfer of the Roman empire by Constantine to Constantinople in 330 CE, the Christian belief that the primary goal in life is gaining salvation and deliverance into heaven replaced attempts to understand and improve the world we live in. If all is ordained by God, gaining God's help by prayer would be more effective in controlling events than discovering their natural causes as in the saying, "Inshallah" or "God willing." As Ambrose, one of the Patristic Fathers and Bishop of Milan, declared: "To discuss the nature and position of the earth does not help us in our hope of the life to come. It is enough to know what Scripture states. . . ."[11] Or as St. Augustine, an early church father and Bishop of Hippo, reiterated: "'Nothing is to be accepted except on the authority of Scripture, since greater is that authority than all powers of the mind.'"[12] But the supremacy of Christianity over paganism began with the zealous Christian Roman ruler Theodosius the Great in 391 CE who issued edicts prohibiting pagan rituals and public ceremonies with the intent of eradicating paganism.

Theophilus of Antioch began applying the edicts of Theodosius directing ruthless gangs of Christians to assault the pagans, along with destroying their sanctuaries, monuments, and statues. Cyril, the nephew and successor of Theophilus, turned the wrath of these Christians against the Jews, ordering their expulsion from Alexandria, but, fortunately, he was opposed by Orestes, governor of Alexandria.

An especially horrific example of the oppressive cruelty was the vile murder of the renown and revered scholar Hypatia. The daughter of a mathematician, she became famous as a mathematician in her own right, along with attaining an outstanding reputation in music, astronomy, and philosophy. Unlike women at the time who where secluded in their homes, she was one of Alexandria's most admired personages: very beautiful as well as learned and refined, she rode around the city in a chariot. But admirable as this reputation was, it led to her vicious execution.

Opposed to her pagan notoriety, in March of 415, upon returning home, a gang of Cyril's followers attached her and took her to a church, where "she was stripped of her clothing, her skin was flayed off with broken bits of pottery. The mob then dragged her corpse outside the city walls and burned it. Their hero Cyril was eventually made a saint."[13] So much for the Christian "brotherhood of man." The period when Christianity was dominant has justly been called the dark ages. In fact, the Inquisition in Spain during the fifteenth century was one of the most unjust, terrifying, and fiendish in history.

Having described the transition from the earlier mythological, theogonic worldview to the first awaking of the possibility of a more rational understanding of the universe and human existence, we turn now to the second revolution when science began replacing both religion and philosophy.

Chapter II

THE SECOND TRANSITION OWING TO THE CREATION OF MODERN CLASSICAL SCIENCE

The second revolution that transformed our conception of reality followed the Renaissance with the resurgence of classical Greek science in the sixteenth and seventeenth centuries and extended to the Age of Enlightenment in the eighteenth century. Realizing the failures of such pseudosciences as alchemy and astrology, it also represented the demise of the authority of the Catholic Church because of the increasing confidence in the new scientific advances disclosing and refuting the incredible nature of Christian beliefs. It also replaced Aristotle's mistaken conception that metaphysics was a separate philosophical discipline that could transcend scientific inquiry by attaining a truer conception of reality that sustained such later philosophical systems as absolute idealism until it was eventually superseded by science.

Though it was mainly Aristotle's insistence that *observable evidence* was the primary basis of natural explanations that originally replaced mythical or religious creation myths, it gradually became clear that sensory evidence alone was insufficient to explain our ordinary experience of natural phenomena and their underlying causes. Three developments especially led to the awareness of the deceptive nature and explanatory limitations of sensory perceptions, along with the necessity of revising and supplementing them with more exact laws and a different conceptual system.

These developments were: (1) more exact astronomical measurements with the astrolabe and sextant; (2) technological advances in observation with the invention of the telescope and the microscope; and (3) the realization that empirical generalizations—such as the Aristotelian law of free fall or Eudoxus' conception of the circular orbits and uniform motion of the planets—must be replaced by more exact observations, experimental investigations if possible, and testable explanations. This culminated in replacing the geocentric perspective with the heliocentric cosmology; the discovery of more exact astronomical and terrestrial laws of motion; and the rejection of the age-old distinction between the celestial and terrestrial worlds with the recognition of a microworld of particles and formulation and adoption of Newton's corpuscular-mechanistic worldview.

This essentially began with Copernicus's magisterial work, *De revolutionibus orbium coelestium* (On the Revolutions of Heavenly Spheres), containing numerous diagrams, mathematical tables, and charts depicting the revolutions of the planets to show the incongruities of the geocentric cosmology and proposing a more harmoniously integrated system. Confirming the resurgence of an interest in Classical Greek science, Copernicus states in his Preface and Dedication to Pope Paul III that since the ancient Pythagoreans had considered that the earth moved,

> when I had meditated upon this lack of certitude in the traditional mathematics concerning the composition of movements of the spheres of the world . . . I finally discovered by the help of long and numerous observations that if the movements of the other wandering stars are correlated with the circular movement of the Earth [around the Sun], and . . . are computed in accordance with the revolution of each planet, not only do all their phenomena follow from that but . . . this correlation binds

> together so closely the order and magnitudes of all the planets and of their spheres or orbital circles and the heavens themselves that nothing can be shifted around in any part of them without disrupting the remaining parts and the universe as a whole.[14] (brackets added)

One can discern one revolutionary feature of his "hypotheses" in the fact that it was "the lack of certitude in the traditional mathematics" that he cites as the justification for attributing motions to the earth despite its opposition to ordinary experience, showing the increasing influence of mathematics in scientific inquiry (an affirmation of Plato over Aristotle later extolled by Galileo). It also shows a new awareness of the tremendous complexity of the problems, along with the required freedom to explore alternative possibilities offered by the new evidence, again rejecting the unwarranted injunctions of the Catholic Church.

Thus began the dismantling of the ancient model of the universe as a celestial sphere composed of an aethereal substance bounded by the sphere of the fixed stars revolving around the stationary central earth. Within this sphere the seven planets, Moon, Mercury, Venus, Sun, Mars, Jupiter, and Saturn, in that ascending order, revolved in heavenly spheres in uniform circular motion at extended distances. Their individual motions were variously attributed to intelligences (Aristotle), to having been created by God in their respective orbits and endowed with their uniform circular motions (Christians), or to their having souls as Kepler originally believed and then replaced by a materialistic "clockwork" system driven by gravity, further emended by Galileo and Newton.

Although Copernicus deserves immense credit for initiating the unraveling considering its consequences, other than exchanging the positions of the earth and the sun he retained most of the explanatory system of the traditional cosmology, such as uniform circular motions. The more radical revisions were by Kepler and Galileo in the seventeenth century and by the telescopic discoveries of Erwin Hubble of the Mount Wilson Observatory in the early twentieth century revealing the existence of billions of additional solar systems, galaxies, and in 1929 the expansion law of the universe.

Kepler was attracted to Copernicus's system—as Copernicus was—because of its harmonious mathematical integration of the astronomical evidence and what Kepler believed would facilitate the discovery of the as yet undetermined exact ratios of the distances, motions, sizes, and of the planetary orbits owing to the sun's influence. Then, when he gained access to Tycho Brahe's naked eye observations (his being the most exact at the time) and the depiction of Mars's orbit noticing its variations in brightness and velocity and that its shape was oval or elliptical, it became apparent that its orbital motion could not be circular and uniform. As he wrote to his friend Johann Fabricius, a clergyman and amateur astronomer in Freesland: "if only the shape were a perfect ellipse all the answers could be found in Archimedes' and Apollonius' works,"[15] especially the latter's conic sections. Illustrating not only his respect for the evidence, it shows the continued influence of the Hellenistic mathematicians. Indicative of an extraordinary flexibility and openness of mind, Kepler discarded his earlier simplistic explanations for a revised final interpretation that even Tycho Brahe was unable to accept.

Also aware that the sun's emission of light became dimmer with the increased distance, it occurred to him that rather than the planetary-souls producing the motion, perhaps another "force" (*vis*) emanated by the sun analogous to light, "corporeal but immaterial," could provide a more effective explanation. This hypothesis was reinforced by his reading of William Gilbert's book, *De Magnete* (On the Magnet), in which he described the earth and other planets, including the sun, as having magnetic poles on their opposite axes. Kepler inferred that if, as the planets revolved around the sun, their respective polarities were opposed causing them to be attracted or repelled depending on their alignment, this would explain the shapes and motions of their orbits deviating from circular and uniform motions. This recourse to forces was unusual at the time.

He thus concluded that the two forces, the sun's emanation causing the planets revolution and Gilbert's contrasting magnetic forces, could explain the observed ovoid or elliptical motion of the planets. But it was not until he measured the deviation of Mar's orbit from circular to elliptical as .00427 of the radius, that he was finally convinced that the orbit must be elliptical. It was this replacement of his earlier explanations, consisting of the Platonic solids and souls, with "forces" that enabled him to formulate his astronomical laws. Noted author Arthur Koestler describes the tremendous import of this change:

> It would be difficult to overestimate the revolutionary significance of this proposal. For the first time since antiquity, an attempt was made not only to *describe* heavenly motions in geometrical terms, but to assign them a *physical cause.* We have arrived at the point where astronomy and physics meet again, after a divorce which lasted for two thousand years. This reunion . . . produced explosive results. It led to Kepler's three Laws, the pillars on which Newton built the modern universe. (p. 258)

While the first of these laws was anticipated in Kepler's early work, *Commentaries on the Movement of Mars,* it was in his *Astronomia Nova* (New Astronomy) published in 1609 (the year Galileo began his telescopic observations), acclaimed as "the first modern treatise on astronomy" that, having read Gilbert's book, he offered his first two planetary laws: (1) "that the planets travel round the sun not in circles but in elliptical orbits," and (2) "that a planet moves in its orbit not at uniform speed but in such a manner that a line drawn from the planet to the sun sweeps over equal areas in equal times" (p. 313).

Yet despite these striking innovations he still had not determined the exact ratios of the velocities of the planets' orbits to their distances from the sun, which was his original motivation in studying astronomy. Initially he assumed the simplest ratio, that the velocity diminished inversely with its distance from the sun, but in the *Dioptrice*, published in 1611, believing that the intensity of light lessened inversely with the *square* of the distance, he attributed this also to the sun's "gravity," a term he had introduced in the *Nova* as the sun's emanation causing the orbital motions.

There he declared that "Gravity is the mutual bodily tendency between cognate [i.e., material] bodies towards unity or contact . . . each approaching the other in proportion to the other's mass."[16] Attributing this gravitational force to all physical objects such as the earth and its moon, he also inferred that it is the earth's attractive force that prevents its seas from rising to the moon. While Galileo rejected this explanation on the basis that it involved an "occult force," it became a crucial principle in Newton's celestial mechanics.

Then in the *Harmonice Mundi* (World Harmony), published in 1619, based on Tycho Brahe's measurements comparing the ratios of the periods of the planets with their distances from the sun, he deduced that "*The squares of the periodic times are to each other as the cubes of the mean distances.*"[17] That is, the period of the revolutions vary with the 3/2th power of their distances: "*it is certain . . . that the ratio which exists between the periodic times of any two planets is precisely the ratio of the 3/2th power of the mean distances, i.e., of the spheres themselves. . . .*"[18] It was this crucial ratio that would provide Newton with the key to his universal law of gravitation.

What an extraordinary achievement! With these laws one could finally cast aside the spiritual or heavenly nature of the universe, along with all the past fabricated devises such as celestial spheres, epicycles, eccentrics, equants, and souls (though not God as the initial cause) previously used to explain the movements and dimensions of the planetary orbits. In fact, replacing the conception of a celestial universe with a mechanistic one seems to have been his ultimate intention, as he wrote to his friend Herwart von Hohenburg, the Catholic chancellor of Bavaria who asked Kepler, along with other astronomers, for his opinion on certain astronomical problems,

> My aim is to show that the heavenly machine is not a kind of divine, live being, but a kind of clockwork . . . insofar as nearly all the manifold motions of the clock are caused by a most simple . . . weight. And I also show how these physical causes are to be given numerical and geometrical expression.[19]

Though this was his vision, in his written works he was unable to present it in such a clear and convincing manner that it lacked the impact it should have had, thus his reputation was largely overshadowed by Galileo and others in the seventeenth century.

Yet his legacy did include a final significant achievement, namely, the creation of the *Rudolphine Tables*. As I wrote previously:

> In 1614 John Napier had published a much praised work, *Merifici Logarithmorum Canonis Descriptio* [Different Descriptions of Logarithmic Canons], containing logarithmic tables that facilitated astronomical calculations, but had not shown how they had been computed. Knowing of its popularity but limited explication, in the years 1621–1622 Kepler wrote a work that contained not only logarithmic tables along with instructions for their use, but also considerable planetary data and a star catalogue comprising over a thousand fixed stars. Published five years later and entitled the *Tabula Rudophinæ* in honor of his deceased patron Rudolph II, it served for over a century as the basis of astronomical calculations and predictions.[20]

I should not leave this discussion without correcting what may have been a false impression that Kepler's intellectual development was a natural, direct, and smooth process and that his personal life was a prosperous one. Nothing could be further from the truth. Not only was his remarkable intellectual development arduous, conflicted, and at times regressive, despite his achievements, his personal life was fraught with tragedy indicative of the times. His first wife and their child died as did three other children from his second marriage; he was beset by constant financial problems due to the Crown withholding his earnings, and he had to face the awful charge of witchcraft brought against his aged mother, a charge that was eventually withdrawn owing to her unwavering insistence on her innocence, even under threat of torture and death, and assisted by Kepler's devoted defense supported by friends at court.

Furthermore, he suffered a final tragedy when, on a trip to Ratisbon, Bavaria, in an effort to regain 11,818 florins owed him by the Crown, he became fatally ill and died three days later on November 15, 1630, apart from his family and alone. As a further misfortune, he was buried in the cemetery of Saint Peter outside of Ratisbon but his actual gravesite is unknown due to the ravages of numerous successive battles there. However, these humiliations are somewhat mitigated by a fine statue of him beside Tycho Brahe on the hill overlooking the city of Prague honoring their outstanding contributions to the creation of modern classical science, especially Kepler's.

Having limited myself to the major contributors to the development of modern classical science, I shall turn now to the outstanding achievements of Galileo comprising his extraordinary, iconoclastic telescopic observations, his experimental discovery of the mathematical law of free fall and proof of parabolic motion, his eloquent endorsement of mathematics as "the language of nature," and his continued dismantling of the competing Aristotelian or Scholastic cosmological system in favor of a mechanistic worldview based on mathematical computations.

It was overhearing a lecture by the court mathematician Ostilio Ricci on Euclid that lured Galileo to the study of mathematics and mechanics. Continuing his tutelage under Ricci studying the works of Eudoxus and Archimedes, Ricci soon recognized Galileo's exceptional mathematical ability and encouraged him to study statics and hydrostatics. This led to Galileo's first scientific publication, *La Bilancetta* (The Little Balance), in 1586, written in Italian that introduced modifications in the Westphal balance enhancing measurements of specific gravity and specific weights.

His next book, *De Motu* (On Motion), described his research leading to his criticism of Aristotle' s theory of terrestrial motions, especially the latter's common-sense law that free-falling objects accelerate in proportion to their weights. In 1602 he published a book on *Mechanics* continuing his incline plane experiments begun in *De Motu* to determine their exact rate of free fall. Like Kepler he tried to determine the ratio of the sizes and speeds of the planetary orbits in relation to the sun, but unlike Kepler was unable to arrive at the correct proportion due to his adherence to circular orbits and his rejection of Kepler's ellipses. But in experiments with pendulums he discovered that when the duration of their swings was equal or isochronal, this was due not to their weights, as generally believed, but to their lengths. This showed Galileo's extensive curiosity about natural phenomena and his usual willingness to challenge traditional beliefs and authority, along with a realization of the importance of experiments in testing what seemed obvious common-sense truths. Still, as the acclaimed historian of science James Gleick points out, as science progresses one discovers that these earlier discoveries are approximate.

> The regularity Galileo saw is only an approximation. The changing angle of the Bob's motion creates a slight nonlinearity in the equations. At low amplitudes, the error is almost nonexistent. But it is there, and it is measurable even in the experiment as crude as the one Galileo describes.[21]

Continuing his incline plane experiments, two years later he confirmed that as the ball rolls from rest down the incline its acceleration in *equal times* traverses distances proportional to the odd numbers beginning with one: 1, 3, 5, 7, 9, etc. In addition, he found that the *square roots* of the *successive sums* of the odd numbers gave the successive times of descent: for example, the sum of 1 and 3 = 4 whose square root is 2; the sum of 1, 3, and 5 equals 9 whose square root is 3, etc. In turn, these numbers squared indicate the ratios of the increases of acceleration during the fall. As he later wrote to Fra Paolo Sarpi whom he greatly admired, "he had found a proof for the square law, the odd number rule, and other things he had long been asserting, if granted the assumption that velocità are proportional to distances from rest."[22] These early experimental discoveries were important because they would become the basis of his second most important book, *Dialogues Concerning Two New Sciences*, published shortly before his death.

Following the sighting in 1604 of a brilliant nova, there was an event in 1609 that would soon transform the entire conception of the universe. Galileo had always exhibited a great curiosity in machines, along with a skill for inventing and using instruments, such as calculators, calibrators, quadrants, even constructing a number of compound microscopes and engaging an instrument maker to live with his family to assist him in his constructions. Thus when occupying the Chair of Mathematics in Venice at age forty-five and hearing a "rumor" of a Dutch instrument called a "spyglass" that brought distant objects closer by magnifying them, it aroused his usual curiosity. As he later wrote in the *Sidereus Nuncius* (The Sidereal Messenger, translated as Starry Messenger):

> About 10 months ago a rumor came to our ears that a spyglass had been made by a . . . Dutchman by means of which visible objects, although far removed from the eye of the observer, were distinctly perceived as though nearby. . . . This . . . caused me to apply myself . . . to investigating the principles and figuring out

> the means by which I might arrive at the invention of a similar instrument, which I achieved shortly afterward on the basis of the science of refraction. And first I prepared a lead tube in whose ends I fitted two glasses, both plane on one side while the other side of one was spherically convex and of the other concave. Then, applying my eye to the concave glass, I saw objects satisfactorily large and close. Indeed, they appeared three times closer and nine times larger than when observed with natural vision only. . . . Finally, sparing no labor or expense, I progressed so far that I constructed for myself an instrument so excellent that things seen through it appear about a thousand times larger and more than thirty times closer than when observed with the natural faculty only.[23]

His first lunar observations were made in December of the same year followed by vivid drawings the following March in the *Sidereus Nuncius* of the phases and irregular surface of the moon showing its pits, mountain ranges, and ravines that he interpreted as evidence of rivers, lakes, and seas. The drawings so clearly resembled the earth that they caused a sensation. Having experienced the videos of astronauts landing on the moon and the recent discoveries of the surface of Mars by the rover Curiosity, it is difficult today to appreciate the shock of those comparisons, since they completely undermined the ancient distinction between the celestial and terrestrial worlds. This was the first *observational evidence* contesting it!

Galileo also included his citing of the "four little stars" circling Jupiter and what are now known as the "rings of Saturn," but that he described as "two opposite protuberances resembling ears," along with the phases of Venus predicted from the heliocentric but not the geocentric system, further opposing the latter. Because it was a defining characteristic of the ethereal realm that it was eternally unchanging, these new appearances also were shocking and threatening to the established cosmology.

Yet perhaps even more astonishing was his telescopic disclosure of the seven stars known as the Pleiades, along with many other fixed stars that had never been sighted with the naked eye. Because they showed no evidence of parallax (indicating the *decreasing* displacement of distantly moving objects the further away they are) implying they were much farther away and that the universe was much more expansive than previously thought. This could even challenge the spherical nature and alleged finitude of the universe. Galileo mentions these implications, but because Giordano Bruno was burned at the stake in 1600 by order of the Holy Office for his heresy in advocating an infinite universe similar to that of Epicurus and adamantly denied the right of the inquisitors to decide what counted as heresy, he did not press the argument.

As these observations obviously conflicted with Aristotle's cosmological view, the latter's followers concocted explanations to discredit the evidence, such as declaring that it was an artifice of the lenses, that the "spyglass" was only effective when used on the earth, or that despite appearances the moon was covered by a transparent crystalline sphere. One is apt to forget how rigidly authority was accepted at the time, both that of the Catholic Church and that of Aristotle. The Aristotelian Cesare Cremonini refused to look through the spyglass on the grounds that the evidence could not be true because it was not mentioned in any of Aristotle's works. Galileo's reply was that it was "as if this great book of the universe had been written to be read by nobody but Aristotle, and his eyes had been destined to see all for posterity."[24]

Undaunted by these arguments and confident of the soundness of his evidence, he wrote in the *Sidereus Nuncius*:

> By oft-repeated observations . . . we have been led to the conclusion that we certainly see the surface of the Moon to be not smooth, even, and perfectly spherical, as the great crowd of philosophers have believed about this and other heavenly bodies,

> but, on the contrary, to be uneven, rough, and crowded with depressions and bulges. And it is like the face of the Earth itself, which is marked here and there with chains of mountains and depths of valleys. (p. 40)

Initially the Catholic Church had supported his telescopic discoveries. In 1611 the prestigious Jesuit School, the Collegio Romano, authenticated his observations and he was inducted into the distinguished Accademia dei Lyncei where, after a banquet in his honor, the name *occhilai* or spyglass was replaced by *telescopium,* according to Stillman Drake. Later, during one of his visits to Rome he was honored by two popes, Pope Paul V assuring him "that he knew of Galileo's integrity and sincerity," and that "so long as he lived . . . Galileo remained secure" (p. 256). Even his successor, Pope Urban VIII (later his livid accuser of duplicity), was such an ardent admirer that when sent a copy of *Il Saggiatore* (The Assayer) he had even written a poem in his honor.

But as his controversial discoveries continued and his disagreements with both the Aristotelians and the ecclesiastical authorities over the interpretation of natural phenomena increased, so did the hostility. For example, when the Aristotelians explained the amount of support water gave to floating objects as due to their shapes, in contrast to Archimedes' principle of specific gravity, Galileo published a reply entitled (in translation), *Discourse on Bodies on or in Water,* supporting Archimedes. In 1612 he entered into a dispute with a Jesuit mathematician named Christopher Scheiner (who wrote under the pseudonym Apelles) over the nature of the recently cited dark spots circling the sun. Scheiner argued that they were tiny stars similar to the four stars circling Jupiter, while Galileo, based on their formation, maintained they were like clouds circling the earth. Again challenging the distinction between the celestial and terrestrial worlds, this proved quite contentious (today sun spots are explained as magnetic fields that emit massive bursts of energy that appear as dark areas on its surface).

Then in a famous "Letter to Castelli" written on December 21, 1613, he openly expressed his disdain for those ecclesiastical authorities who rejected his discoveries because they conflicted with traditional biblical beliefs. Conceding that regarding questions concerning salvation and faith there was no higher authority than Holy Scripture, he adds that

> I should think it would be prudent if no one were permitted to oblige Scripture . . . to sustain as true some physical conclusions of which sense and demonstration and necessary reasons may show the contrary. . . . I do not think it is necessary to believe that the same God who has given us our senses, reason, and intelligence wished us to abandon their use, giving us by some other means the information that we could gain through them. . . . (p. 226)

Though a rationally sound objection, the clergy considered it not only as a rejection of the heavenly nature of the universe, but also as contesting the authority of scripture and the Church itself, a crucial turning point in his relation with the Church. The following year on December 21 a fiery young Dominican named Tommaso Caccini "denounced from the pulpit of Santa Maria Novella the Galileists, and all mathematicians along with them, as practitioners of diabolical arts and enemies of true religion" (p. 238). About the same time, the cardinals of the Inquisition started examining Galileo's writings to see if they contained heretical material. Hoping to defend himself, he journeyed to Rome at the end of 1615 but with little success.

A commission of theologians was formed in February of 1616 that decided against the motions of the earth and the centrality of the sun, instructing Cardinal Bellarmine to inform Galileo of its decision, after which he was told to abandon those suppositions. Bellarmine met with Galileo on February 24 before a notary and a witnesses, leaving a notarized but unsigned record stating, in the words of Drake, that he "told Galileo of the official findings against the motion of the earth and stability of the sun," while the commissary of the Inquisition "admonished Galileo in the name of the pope that he must not hold, defend, or teach in any way, orally or in writing, the said propositions on pain of imprisonment. Galileo Agreed" (p. 253). This is crucial in connection with his final trial and conviction in that Galileo *did agree* "not to hold, defend, or teach in any way, orally or in writing the motion of the earth and stability of the sun." An edict was then dispensed proscribing all books purporting to reconcile Christianity with heliocentrism, though none of Galileo's were included. Finally, in 1992, the Catholic Church acknowledged that Galileo was correct and it was wrong.

Then in the fall of 1618 the citing of three comets again evoked the question of the reality of the distinction between the celestial and terrestrial worlds depending on the location of the comets, the distinction that previously had been raised by Galileo's lunar observations. Orazio Grassi (writing under the pseudonym of Lothario Sarsi of Siguenza), a well-known astronomer and critic of Galileo's observations, argued that because there was no evidence of parallax (again no change in the position of the stars as one moved) nor of enlargement, they must be in the translunar world and thus should have caused no opposition on the part of Galileo. But because Grassi's (or Sarsi's) argument embraced Tycho Brahe's modified geocentric view that the sun, encircled by the planets, revolved around the central Earth, Galileo dismissed it because of its asymmetry. In his rebuttal he not only ridiculed Tycho's system, he also mocked the distinction between the two worlds so cherished by the Aristotelians and the Christians, declaring "[n]ever having given any place in my thoughts to the vain distinction (or rather contradiction) between the [terrestrial] elements and the heavens . . ."[25] (brackets added).

His second reply to Grassi in the *Il Saggiatore* (or The Assayer) written in 1623, is extremely important because it contains a further crucial revision of the traditional worldview. Drawing a sharp distinction between the ordinary sensory world and the independent micro-mechanistic world whose particles, being devoid of sensory qualities, were defined in terms of measurable physical properties, such as mass, motion, shape, and size, this would greatly contribute to the transition to Newton's corpuscular-mechanistic cosmology whose reality and exact nature posed the central problem of science and philosophy during the following three centuries.

I know of no previous or even later analysis to match Galileo's meticulous justification of the distinction, in section XLVIII of *Il Saggiatore*, by analyzing the nature and origin of sensory qualities. While we normally distinguish pains and tickling sensations as being obviously subjective, we think of colors, sound, tastes, hardness, and heat as residing in the objects surrounding us independently of their being perceived. But having learned more about how dependent these latter sensory experiences are on our sense organs, nervous system, and the brain, Galileo argued that they too should be considered as subjective. But as previously indicated, how neurophysiological processes in the brain create the perceptual world as we experience it is still one of the greatest (if not *the* greatest) mysteries confronting us.

Both scientists and philosophers talk as if the ordinary perceptual world, being dependent on our brains, exists in our brains. But does that really make much sense, any more than saying it exists in the pineal gland, as René Descartes held? Certainly the Apple® computer I am using in composing this does not simply exist in my brain, nor does the car I get into and drive, the apartment I live in, the wife I live with, or the body I have. If my body exists in my brain because it is perceived, then since my brain is part of my body it, too, must exist in the brain, which does not make sense. Does the pistol someone uses to commit suicide exist in their brain? Did the nuclear disasters in Hiroshima and Nagasaki merely exist in people's brains?

Does it not make more sense to acknowledge that the world in which we exist, which includes colors, sounds, tastes, etc., is objectively real *within the conditions in which we experience it*, which seems to be true of the various dimensional contents of the universe as a whole? This does not preclude the necessity of revising our conception of this world, as in the Copernican revolution, but of recognizing the conditional status of all that is experienced and exists. This is the thesis I will be defending: that the universe consists of a seemingly endless series of objective contexts or conditions that is the destiny of scientists to explore and understand.

Galileo's justification of the distinction between the independent external causes of these sensory experiences and their modifications or additions due to their interaction with the human organism is clearly described. Though he discusses each sensory quality individually, I think the clearest general statement of his position is the following:

> I do not believe that for exciting in us tastes, odors, and sounds, there are required in external bodies anything but sizes, shapes, numbers, and slow or fast movements; and I think that if ears, tongues, and noses were taken away, shapes and numbers and motions would remain but not odors or tastes or sounds. These, I believe, are nothing but names, apart from the living animal—just as tickling and titillation are nothing but names when armpits and the skin around the nose are absent.[26]

Since nothing was then known about the molecular, atomic, or subatomic structure of matter, he assigned the sizes, shapes, numbers, and movements to the "minute particles" or "corpuscles" that he believed constituted material objects. As examples, he says sounds "are created and are heard by us when . . . a rapid tremor of the air, ruffled into very minute waves, moves certain cartilages of a tympanum within our ear . . . that vision, the sense which is eminent above all others, is related to light . . . and that a multitude of minute particles having certain shapes and moving with certain velocities" striking the senses produce "the sensation which we call *heat*" (pp. 311–12).

These independent physical properties, later named "primary qualities," versus the subjective "sensory qualities," by John Locke in his *Essay Concerning Human Understanding*, became accepted scientific distinctions constituting Newton's corpuscular-mechanistic worldview. But the attempt to discover the actual nature of these particles and corpuscles and their properties, along with how they produce the sensory effects they do has been a major challenge of scientific research ever since. Thus it is fair to say that Galileo helped set the agenda of the physicists, chemists, physiologists, and microbiologists of modern science, along with the epistemological problems of Descartes and Locke, as well as most twentieth-century philosophers.

Galileo himself was aware of his enormous originality and influence, immodestly listing his various books and their contributions in a letter to Belisario Vinta, a close scientific friend, seeking a better position. As again quoted by Drake, they consist of

> two books on the system and constitution of the universe—an immense conception full of philosophy, astronomy, and geometry; three books on local motion, an entirely new science, no one else, ancient or modern, having discovered some of the very many admirable properties that I demonstrate to exist in natural and forced motions, whence I may reasonably call this a new science discovered by me from its first principles: three books on mechanics . . . and though others have written on this same material, what has been written to date is not one-quarter of what I write, either in bulk or otherwise. (p. 160)

Though not mentioned, he also asserted the valid principles of the uniformity of nature and the conservation of momentum.

This brings us to his most famous book, the English title of which is *Dialogue Concerning the Two Chief World Systems—Ptolemaic & Copernican*, whose renown is based on two factors: (1) it's astute arguments written in colloquial Italian and presented in dialogue form essentially showing the superiority of the heliocentric cosmology that made it the greatest scientific dialogue ever written; and (2) the scandalous conviction of Galileo of heresy by the Catholic Inquisition based on his alleged duplicity in writing the book, that has been called "the disgrace of the century."

Having described in another work the contents of his book and the ensuing trial and conviction in greater detail, I shall focus mainly on the arguments he introduced to refute the objections to the movements of the earth. Taking place over four days, the dialogue is between three interlocutors, one of whom is Salviati, a Florentine aristocratic friend who has the role of an academician representing Galileo; another is Sagredo, a Venetian nobleman who acts as the moderator; and the third named Simplicius after a sixth-century scholastic who defends Aristotelianism and the connotations of whose name perhaps added to his selection as the opponent.

The first day's dispute concerns the distinction between the sublunar and translunar worlds. Salviati argues that while the distinction may have been warranted in Aristotle's day, new telescopic evidence such as Galileo's observations showing the moon's surface to be similar to the earth's; the discovery of four stars circulating Jupiter; the rectilinear trajectory of meteors; the detection of sun spots, novas, and new stars; along with Kepler's discovery of the elliptical shape of Mars's orbit is strong evidence that the distinction is no longer valid.

The second day begins with Simplicius defending, as was commonly believed at the time, the complete authority of Aristotle's writings: "There is no doubt that whoever has this skill will be able to draw from his books demonstrations of all that can be known; for every single thing is in them."[27] But previously Galileo had Sagredo express his firm belief in the limits, at the time, of knowledge: "there is not a single effect in nature, even the least that exists, such that the most ingenious theorists can arrive at a complete understanding of it" (p. 101).

But turning to the main dialogue of the day, as Galileo had argued, rather than accept the usual explanation that the apparent rising and setting of the sun was caused by the entire universe revolving from east to west in a single day, Salviati points out that the same appearance could be explained more simply and harmoniously by attributing a diurnal rotation to the much smaller earth from west to east. Not only was it incongruous to have the sphere of the fixed stars, at the farthest distance from the earth, complete their revolution in one day while the closer planets completed theirs in a much longer time, the westward revolution attributed to the fixed stars was contrary to the eastward revolution of the planets. Thus Salviati concludes that "by making the earth itself move, the contrariety of motions is removed, and the single motion from west to east accommodates all the observations and satisfies them all completely" (p. 117).

Salviati then addresses the counter argument that despite its simplicity, attributing the diurnal rotation to the earth cannot be true because then one would see clouds, birds, or other aerial objects displaced to the west as the earth revolved eastward. The example of dropping an object from the masthead of a ship is introduced, declaring that during the fall the object would descend at an angle inclined further from the masthead rather than parallel to it, even affirming that the experiment had been performed and shown the described result.

Salviati replies that this could not be true because when he had performed the experiment a solid object dropped from the masthead of a *uniformly moving* ship fell parallel to the masthead. He reinforces his argument by pointing out that in the cabin of a uniformly moving ship (as in an airplane today) everything happens as if the ship were stationary; objects dropped or thrown have the same trajectory as if the cabin were at rest. Since the objects partake of two motions, that of the uniformly moving container and the downward fall, the former cancels out leaving only the falling object as visible. Thus an object dropped from a tower falls parallel to the tower despite the rotation of the earth during the fall. As Salviati concludes:

> With respect to the earth, the tower, and ourselves, all of which keep moving with the diurnal motion along with the stone, the diurnal movement is as if it did not exist; it remains insensible, imperceptible, and without any effect whatever. All that remains observable is the motion which we lack, and that is the grazing drop to the base of the tower. (p. 171)

While the second day's dialogue addressed the opposition to the earth's rotary motion, the third day deals with Simplicius' dissent to the earth's annual revolution around the sun based on the ordinary experience of seeing the sun circle the earth and the fact that as a terrestrial heavy body the earth naturally should be in the center of the cosmos. Salviati first responds by pointing out that Simplicius' argument presupposes that the cosmos is a finite sphere, yet it has not been proven whether that is its shape or whether it is "infinite and unbounded." But as indicated previously, because Giordano Bruno was burned at the stake by order of the Holy Office partially for advocating an infinite universe, this argument is not pursued further.

Instead, Salviati offers Kepler's second two astronomical laws based on the sun's gravitational force as evidence of its central position and Galileo's telescopic evidence of Mars's radical deviation from a circular orbit, as seen from the earth, as refuting the Aristotelian view that all the planetary orbits are circular. As Salviati states regarding several orbital trajectories:

> This approach and recession is of such moment that Mars when close looks sixty times as large as when it is most distant. Next,

> it is certain that Venus and Mercury must revolve around the sun, because of their never moving far away from it, and because of their being seen now beyond it and now on this side of it, as Venus's changes of shape conclusively prove. (p. 322)

He next cites Galileo's telescopic observations showing that the obits of Mercury and Venus are below Earth's while those of Mars, Jupiter, and Saturn are above it. When Simplicius refers to the anomaly in the Copernican system of only the Moon revolving around Earth while all the other planets revolve around the Sun, Salviati replies that this anomaly has been mitigated by the discovery of the four satellites circling Jupiter. Sagredo then brings up two other objections, the observed retrograde, loop-like motion of the five planets as seen from the earth and the absence of parallax when observing the stars. As for the first, Sagredo refers to a diagram by Galileo showing that "these stoppings, retrograde motions, and advances," are illusions produced by the annual revolution of the earth around the sun (p. 342).

Regarding the absence of parallax or displacement when viewing the stars from the different positions on the earth as it revolves around the sun, this can be explained by attributing a much greater distance to the stars than normally believed. The Aristotelian response was that for a star to be seen from such a great distance "it would have to be so immense in bulk as to exceed the earth's orbit—a thing which is, as they say, entirely unbelievable" (p. 372). Lacking any evidential rebuttal, Salviati gives the sensible answer that without knowing how the stars transmit their light from such a great distance it is impossible to draw a definite conclusion.

The day's dialogue ends with a summary of the evidence that Salviati (as Galileo) believes shows the greater credibility of the heliocentric system, which is one of the major reasons Pope Urban VIII later felt so strongly that he had been deceived and disobeyed when he had agreed to the publication, as long as Galileo treated both systems impartially.

> See, then, how two simple noncontradictory motions assigned to the earth, performed in periods well suited to their sizes, and also conducted from west to east as in the case of all movable world bodies, supply adequate causes for all the visible phenomena. These phenomena can be reconciled with a fixed earth only by renouncing all the symmetry that is seen among the speeds and sizes of moving bodies, and attributing an inconceivable velocity to an enormous sphere beyond all the others, while lesser spheres move very slowly. Besides, one must make the motion of the former contrary to that of the latter, and to increase the improbability, must have the highest sphere [to] transport all the lower ones opposite to their own inclination. I leave it to your judgment which has the more likelihood in it. (p. 396)

Considering the acuteness of the arguments of the first three days, the fourth and last day's dialogue is disappointing and yet Galileo considered it one of the most convincing evidences of the earth's motions, even to the extent of intending to include it in the title of the book until prevented from doing so by Pope Urban VIII, who thought it presumptuous. Recall that Kepler had explained the tides as caused by the mutual gravitational force between the earth and the moon which was accepted by the scientific community.

Objecting that the explanation invoked a mysterious force acting at a distance similar to "occult powers," Galileo dismissed it. Instead, he attributed the ebb and flow of the waves to the contrasting motions of the earth, similar to the undulations in the water in the bilge of a ship due to its pitching in rough seas. Thus he has Sagredo, the moderator, include this argument with the others in a concluding statement.

> In the conversations of these four days we have, then, strong evidences in favor of the Copernican system, among which three have been shown to be very convincing—those taken from the stoppings and retrograde motions of the planets, and their

> approaches toward and recessions from the earth; second, from the revolution of the sun upon itself, and from what is to be observed in the sunspots; and third, from the ebbing and flowing of the ocean tides. (p. 462)

When published in February 1632, after two years of difficult negotiations, as was to be expected, the contrasting reactions were striking. As reported by Drake, Castelli, who was mentioned previously as his scientific friend and who had been sent a copy, replied: "I still have it by me, having read it from cover to cover to my infinite amazement and to my delight; and I read parts of it to friends of good taste to their marvel and always more delight, more to my amazement, and with always more profit to myself."[28]

The reactions of the clergy, especially the Jesuits and Pope Urban VIII, were not only vivid, but livid. While the consequent trial of Galileo involved considerable misunderstandings and incriminations, that Galileo had previously "agreed" to the edict of February 26, 1616, "that he must not hold, defend, or teach in any way, orally or in writing," the "motion of the earth and stability of the sun," and did not inform the pope of this when he obtained permission from him to published the book, was one of the incriminating charges. As mentioned previously, the other was that the pope, having agreed to the publication provided Galileo "treat the evidence impartially," yet having Salviati declare the "improbability" of the Aristotelian view in contrast to the Copernican system and having Sagredo state "in the conversations of these four days we have, then, strong evidence in favor of the Copernican system," the pope decided that Galileo had deliberately disobeyed his admonition to treat them impartially.

As a result "on 22 June the sentence of life imprisonment was read to Galileo at a formal ceremony in the presence of the cardinals of the Inquisition and witnesses, after which he had to abjure on his knees before them" (Drake, p. 351). Drake adds that "three cardinals of the ten refused to sign the sentence," while Cardinal

Francesco Barberini, the pope's nephew who had aided Galileo throughout the trial, "immediately commuted the place of Galileo's imprisonment at Rome to the Florentine embassy there" (pp. 351–52). Francesco Niccolini, the Tuscan ambassador to Rome, "then undertook to secure a pardon for Galileo from the pope, who refused, but permitted Galileo to be moved to the custody of Archbishop Ascanio Piccolomini, of Siena" (pp. 351–52). Near the end of 1633 he was permitted to be "imprisoned" in his villa at Arcetri where he remained for the rest of his life.

It was there in the few years remaining to him that he wrote *Dialogues Concerning Two New Sciences*, his second most famous work, published in 1638. Retaining the same three disputants the four days of dialogue recounts much of his earlier research on motion and on falling objects, but declares "his purpose is to set forth a very new science dealing with a very ancient subject."[29] He then provides an excellent statement of the scope and originality of his book:

> There is, in nature, perhaps nothing older than motion, concerning which the books written by philosophers are neither few nor small; nevertheless I have discovered by experiment some properties of it which are worth knowing and which have not hitherto been either observed or demonstrated. Some superficial observations have been made, as, for instance, that the free motion of a heavy falling body is continuously accelerated; but to just what extent this acceleration occurs has not yet been announced; for so far as I know, no one has yet pointed out that the distances traversed, during equal intervals of time, by a body falling from rest, stand to one another in the same ratio as the odd numbers beginning with unity. (p. 147)

Galileo was unaware apparently that the odd number law, which he had experimentally proven in 1604, had been previously formulated by Nicole Oreme in the fourteenth century.

He continues by describing some additional contributions that he foresees as just the beginning of a whole new world of discoveries.

> It has been observed that missiles and projectiles describe a curved path of some sort; however no one has pointed out the fact that this path is a parabola. But this and other facts, not few in number or less worth knowing, I have succeeded in proving; and what I consider more important, there have been opened up to this vast and most excellent science, of which my work is merely the beginning, ways and means by which other minds more acute than mine will explore its remote corners. (pp. 147–48)

Rejecting Aristotle's explanation of projectile motion as requiring a contiguous, continuous mover, he seems to have adopted Jean Buridan's fourteenth-century explanation that motion is caused by the mover impressing a force or "impetus" on the projectile that continues the motion without the presence of the mover, though he did not anticipate the law of inertia because of the influence of gravity. Some of these contributions were not discoveries of specific laws but refinements of scientific methodology, such as beginning with the simpler aspects of a problem before moving to more complex ones.

He antedates Newton's formal method of presentation by dividing his analysis of motion into "Definitions, Axioms, Theorems, and Propositions." Beginning with the simplest, uniform motion, and then moving to accelerated and unnatural or projectile motions, this is followed by four axioms and six theorems with diagrams to illustrate his reasoning. His method anticipates that of modern science in rejecting a priori or even common-sense definitions in favor of those best "fitting natural phenomena" *supported by experimental evidence.* The clearest example was his incline plane experiments to demonstrate the accelerated law of free fall, providing an exceedingly detailed and precise description of the experiment.

> A piece of wooden molding or scantling, about 12 cubits long, half a cubit wide, and three finger-breadths thick, was taken; on its edge was cut a channel a little more than one finger in breadth; having made this groove very straight, smooth, and polished, and having lined it with parchment . . . we rolled along it a hard, smooth and very round bronze ball. Having placed this board in a sloping position, by lifting one end some one or two cubits above the other, we rolled the ball . . . along the channel, noting . . . the time required to make the descent. We repeated this experiment more than once in order to measure the time with an accuracy such that the deviation between two observations never exceeded one-tenth of a pulse-beat. (p. 171)

Incredibly, despite this exact description, two of Galileo's contemporaries, Maria Mersenne and René Descartes, questioned their authenticity, Mersenne declaring "I doubt whether Galileo actually performed the experiments of fall down incline planes, since he does not speak of them, and since the ratio he gives is often contradicted by experiment." Descartes "'denied' all of Galileo's experiments! Because . . . those . . . which resulted in measurements, in precise values, were falsified by his contemporaries."[30] But given the meticulous description of his experiments, I do not see in fairness how anyone could deny that he performed them and since his ratio is correct, if his contemporaries found them false, it was because they were incompetent, not that he was!

Following the publication of his book Galileo lived four more years, blind and exhausted, dying on January 9, 1642, less than two months before his seventy-eighth birthday. His body was "privately deposited" in the magnificent church of Santa Croce in Florence. The Grand Duke intended to honor him with a splendid tomb similar to that of Michelangelo, but was prevented by the Catholic Church that "forbade any honors to a man who had died under vehement suspicion of heresy" (Drake, p. 436). Now, as is befitting, there does exist a sepulcher opposite that of Michelangelo and just as grand as he deserves. For as Maurice Clavelin's summary of his contributions justly states:

> The reason, therefore, why no scientific problem was ever the same again as it had been before Galileo tackled it lay largely in his redefinition of scientific intelligibility and in the means by which he achieved it: only a new explanatory ideal and an unprecedented skill in combining reason with observation could have changed natural philosophy in so radical a way. No wonder then that, as we read his works, we are struck above all by the remarkable way in which he impressed the features of classical science upon a 2000-year-old picture of scientific rationality.[31]

There is no way I could add to such a deserving and splendid tribute.

Chapter III
THE CULMINATING ACHIEVEMENT OF NEWTON

In a previous quotation Galileo had stated that his "work is merely the beginning . . . by which other minds more acute than mine will explore its remote corners" that provides an excellent transition to Newton. As we shall find, Newton's astronomical explanations will reflect Kepler's three laws while his conception of the proper scientific method and investigations of terrestrial motions will follow the initiatives of Galileo. The resemblance between his statements about scientific inquiry are strikingly similar to Galileo's, justifying his gracious acknowledgement that his achievements were possible because of those who came before him.

Newton has acquired the greater reputation for having combined the discoveries and laws of Kepler and Galileo into a unified system of laws within a theoretical framework known as the corpuscular-mechanistic worldview with its absolute space, time, and motion that guided scientific research during the following two centuries. Yet that it was made possible by his predecessors makes one wonder who should get the most credit, those who initiated and created the foundations of an entirely new scientific framework in face of much opposition, or those who completed the task based on the earlier discoveries? That is, which is the greater challenge and deserves more credit, breaking with an ancient and entrenched conceptual and cultural tradition and laying the foundations for a new one, or creating a more unified and advanced system based on the previous discoveries and innovative laws and theories of one's predecessors?

Recall that Kepler not only formulated the first exact astronomical laws that refuted the traditional conception of uniform circular motions, he explained them based on his final theory that it was the gravitational force from the central sun that produced these laws, thereby rejecting and replacing his earlier traditional view that it was the planets' souls that caused the motions. In addition, as previously quoted, he introduced the revolutionary conception "that the heavenly machine is not a kind of divine, live being, but a kind of clockwork" and showed how its "physical causes are to be given numerical and geometrical expression." It was these original conceptual advances that helped inaugurate the succeeding extensive cosmological changes.

Then there were the equally revolutionary contributions of Galileo: his astonishing telescopic evidence of the similarity of the Moon's surface (previously considered a celestial body) to that of the terrestrial earth, along with his additional discovery of new astronomical bodies and movements in the previously revered heavenly world that refuted the primordial distinction between the two realms. His inclined plane experiments proving that the acceleration of free-falling objects did not depend on their weights, as Aristotle had claimed, but on the increments of time, and from these experiments he inferred that once in motion a moving body apart from any further causes or resistance of its motion would continue to remain at rest or *move in a straight line* without further reinforcing causes, as expressed in Newton's first two laws of motion.

Equally important was his crucial distinction between the qualitative perceptual world and the microworld of insensible particles influenced by his microscope observations (analogous to his telescope discoveries) that disclosed a vast world of moving particles that he defined in terms of their measurable physical properties. It was this discovery, along with his famous declaration that "the language of nature is mathematics," that became the theoretical foundation of the corpuscular-mechanistic worldview. Adopted, extolled, and extended by Newton, this explains the striking similarity between the statements of Galileo and Newton, which even today is not always recognized, though Newton himself was exceedingly generous in acknowledging his indebtedness to both Kepler and Galileo, along with others.

Finally, both Kepler and Galileo contributed to the replacement of Aristotle's deductive and biological method of scientific explanation based on forms, essences, species, genera, and final causes with the image of the machine consisting of moving particles and forces depicted by mathematical laws. Thus despite Aristotle's acute biological investigations and conceptions that were still prevalent at the time, even biologists began viewing the animal body as similar to a machine, exemplified in the writings of Andreas Vesalius in the sixteenth century and William Harvey in the seventeenth.

In stressing the considerable influence Kepler and Galileo had on Newton I do not mean to detract from the significance of his own explanation of motion, formulation of his two famous "Axioms, or Laws of Motion" (stated in the *Principia,* Bk. I), and his universal law of gravitation "Proposition LXXVI" (also in the *Principia,* Bk. I) influenced by their discoveries that provided the basis of scientific research in the following two centuries until the introduction of Max Planck's quantum mechanics and Albert Einstein's theories of relativity, but to emphasize the importance of the initial contributions of Kepler and Galileo, which I believe is often unrecognized.

Turning now to Newton, despite my previous comments I, along with others, consider him the greatest scientist because of his outstanding contributions to the three major areas of scientific inquiry: *experimental* based on his prismatic discovery that white light is composed of discrete rays of colors which he interpreted as corpuscular; *theoretical* because of his unification of the universal laws of motion and system of celestial mechanics; and *mathematical* owing to his exceptional ability in creating fluxional (differential) calculus to calculate the dimensions of the planetary orbits and in applying it to empirical relations or functions in general. While there are many scientists who are exemplary in one or two of these areas, it is highly unusual to excel in all three.

During a visit to the Royal Society in London in 2007 the head of the Library and Information Services, Keith Moore, asked if there was a particular scientist I was interested in, and when I answered Newton he graciously showed me the first edition of Newton's *Principia Mathematica,* a large handwritten tome with marginal notes. Then, when I told him that I thought Newton was the greatest scientist, he told me that when the members of the Royal Society considered the same question recently they, too, voted for Newton (over Einstein) by a small margin—but one can't help wondering what the vote would have been had it been taken in Berlin.

Like Kepler and Galileo, Newton had an unusual background, along with a peculiar personality. He was born in Woolsthorpe, Lincolnshire, on Christmas Day in 1642, the year Galileo died—an auspicious coincidence—though the events that immediately followed were not. Three months before his birth his father died and three years later his mother remarried. Deciding to live with her new husband in his home, she left Newton in the family manor house in Woolsthorpe with his grandmother, whom he was not fond of. When his mother's second husband died ten years later, she returned to the family home with two children by her second marriage.

Whether it was this unfortunate early beginning or his homosexuality (as evinced by two nervous "disorders" due to the termination of two close male relationships) or both, he was an extremely sensitive, serious, and withdrawn person who shunned controversy and publicity to the extent that he refused to publish some of his articles to avoid disputes. His early education at the Free Grammar School of King Edward VI consisted of traditional religious studies, along with courses in Greek and Latin. Having shown considerable intellectual promise, the headmaster of the school urged his maternal uncle to have him take the necessary preparatory courses for admission to a college or university. Completing these, in the summer of 1661 he enrolled in Trinity College, Cambridge, where his uncle had attended.

At Trinity he pursued his former courses along with studying Aristotle's physics and cosmology, though he soon became more interested in the scientific works of Kepler, Galileo, Pierre Gassendi, Descartes, Robert Boyle, and Henry More. There being more interest in Cambridge University at the time in Descartes's optical research and theory of vortices than in Galileo's discoveries, Newton began studying Descartes; but finding shortcomings in his explanations of light and colors, as well as in his theory of vortices, because it did not explain eclipses or agree with Kepler's three laws, Newton presented his criticisms in a work entitled "Questiones" published in 1664. That was the beginning of his scientific career. He then became interested in the corpuscular-mechanistic theory introduced by Galileo and developed by Gassendi and Boyle. In addition, he began his mathematical studies that showed his brilliance as a mathematician that would prove so crucial to his scientific explanations.

He was elected to a fellowship at Trinity in 1664 at the age of twenty-two, and during the following two years of intense study, referred to as his "*anni mirabilis*" (or miracle years, foreshadowing Einstein), applying his newly acquired mathematical skills, he was able to write three original papers on the problem of motion. In one titled "The Laws of Motion" he made two theoretical discoveries that later were included in his system of dynamics presented in his great book, *Philosophiæ Naturalis Principia Mathematica.* In another he rejected Descartes's theory of vortices because it depended on direct contact to explain mutually reciprocating forces. In the third he introduced the principle of the conservation of momentum in mechanics.

Then, as a result of his critique of Descartes's theory of vortices, he began applying mathematics to the study of planetary motions. As he wrote in a famous letter to William Stukeley, another of his friends who was a fellow student at the grammar school in Grantham (which I was also shown during my visit to the Royal Society), it was while he was at home because of the closing of Cambridge University during the plague years of 1665–1666 that, when sitting in the orchard and seeing an apple fall, it occurred to him that the same gravitational force that caused the apple's fall could produce the elliptical deviation in Mars's orbit if extended that far. Thus the account is factual, not fictional, as sometimes alleged.

Moreover, having read Kepler's works at the time of his mathematical studies, this insight may have been reinforced by Kepler's explanation of the tides as being due to the mutual gravitational force between the earth and the moon. And since Kepler's third law states that a planet's periodic time is proportional to the $3/2^{th}$ power of its mean distance from the sun, it might have occurred to him that the strength of the earth's gravitational force on the moon could be in the same ratio. As quoted by Richard S. Westfall in his superb *A Biography of Isaac Newton*:

> In the beginning of the year 1665 . . . I began to think of gravity extending to y^e orb of the Moon & . . . from Kepler's rule of the periodical times of the Planets being in sesquialterate [$3/2^{th}$] proportion of their distances from the center of their Orbs, I deduced that the forces wch keep the Planets in their Orbs must [be] reciprocally as the squares of their distances from the centers about wch they revolve: & thereby compared the force requisite to keep the

> Moon in her Orb with the force of gravity at the surface of the earth, & found them answer pretty nearly. All this was in the two plague years of 1665–1666. For in those days I was in the prime of my age for invention & minded Mathematicks and Philosophy more than at any time since."[32] (brackets added)

As Kepler had proposed in the *Epitome Astronomæ Copernicanæ*, the laws that he originally applied only to Mars refer to all of the planets, including the moon and the satellites of Jupiter. This could have been a further influence on Newton.

In addition to these investigations and discoveries, he began his experiments on light being dissatisfied with the current theories of color proposed by Descartes, Boyle, and Robert Hooke. It was then thought that light was homogeneous, the different colors produced by it being refracted when striking the retina. Red and blue were considered the dominant colors arising from the greatest refraction of light while the other colors were inadequate when Newton began his prismatic experiments that displayed his remarkable experimental ingenuity complementing his mathematical expertise.

Discovering that ordinary light when refracted through a prism disperses into a spectrum of particular colors that he called "rays," he surmised that the retina acts as a prism refracting ordinary light into the various distinct colors. This was confirmed when he redirected the dispersed rays into another prism and they again became blended. He also concluded that this would explain the rainbow that had puzzled scientists throughout the ages and noticed that the circular lenses he used to measure the dimensions of the colors produced rings of color now called "Newton's rings."

Believing the corpuscular-mechanistic theory to be the basic infrastructure of the universe, this undoubtedly influenced his interpretation of light rays as corpuscular, along with the discovery that the sharp edges of shadows supported this interpretation, despite such scientists as Hooke and Christiaan Huygens defending the wave theory. As Westfall states: "No other investigation of the seventeenth century better reveals the power of experimental inquiry animated by a powerful imagination and controlled by rigorous logic" (p. 164). Later Newton would encounter evidence of the wave theory but persist in defending the view that the rays of light are corpuscular.

Not until the nineteenth century would there be evidence that light is an electromagnetic radiation with wave properties, and it was not until Einstein's explanation in 1905 of blackbody radiation as caused by light consisting of discrete units of energy later called "photons," that the disconcerting notion of the dual nature of light as "wavicles" would be introduced, either property depending on the experimental conditions.

Despite these scholarly accomplishments, the criteria then for promotion to a fellowship—a precondition at Cambridge University for attaining a permanent university position—being based mainly on social status, patronage, and strong academic affiliations that Newton lacked, his prospects were not encouraging. This was increased by his being examined on Euclid by Isaac Barrow, the Lucasian Professor of Mathematics at Cambridge, in connection with the fellowship, and his belief that he had not performed well because at the time he had been studying Descartes's geometry rather than Euclid's. Nonetheless, for whatever reason, he was elected a Minor Fellow on October 1, 1667, and as a Major Fellow nine months later.

But several years later, when he showed Barrow his method for calculating an infinite series, this so impressed Barrow that he sent a copy to John Collins, one of the outstanding mathematicians at the time and disseminator in Europe of unusual mathematical developments, describing Newton as "a fellow of our College, & very young . . . but of an extraordinary genius and proficiency in these things" (p. 202). This was most fortunate because there was no one better positioned to evaluate Newton's mathematical achievements and promote his reputation. Collins later sent him complex problems that he would solve in a short time and return. Impressed, Collins distributed copies to other mathematicians in England, Scotland, and Europe, but terminated the correspondence when he encountered resistance from Newton, who was adverse to publicity and possible disputes.

Nonetheless, the initial correspondence enhanced Newton's reputation, which may explain why Barrow resigned his Lucasian Professorship and recommended Newton as his successor following Newton's appointment as a Major Fellow. Thus, at the early age of twenty-seven he attained the very lucrative and prestigious chair of Lucasian Professor. But though the appointment to the status of Major Fellow "would follow automatically when he was created Master of Arts nine months hence," it was not without its adverse conditions. With two exceptions, the recipients "were required to take holy orders in the Anglican church within seven years of incepting M.A." And so on October 1667 Newton

> became a fellow of the College of the Holy and Undivided Trinity when he swore "that I will embrace the true religion of Christ with all my soul . . . and also that I will either set Theology as the object of my studies and will take holy orders when the time prescribed by these statues arrives, or I will resign from the college." (p. 179)

To fulfill the conditions he would have had to "set theology as the object of my studies and . . . take holy orders," which meant remaining celibate and embracing the Athanasian Creed. While remaining celibate was not a problem for personal reasons mentioned earlier, nor was studying theology since he had spent a number of years researching early church history. But swearing to adhere to the Athanasian Creed was a problem. As a result of his extended studies he had decided that the Council of Nicaea's decision to adopt the Trinitarian doctrine of the Athanasian Creed that

God, Jesus, and the Holy Spirit were "consubstantial" was a "massive fraud, which . . . had perverted the legacy of the early church" (p. 313). Thus he embraced the Arian Creed that Jesus and the Holy Spirit had been created by God, not consubstantial with Him.

Fortunately, to prevent his having to resign his fellowship along with the Lucasian Chair at Trinity, an official dispensation was passed on April 27, 1675: "By its terms, the Lucasian professor was exempted from taking holy orders unless 'he himself desires to . . .'" (p. 333). Furthermore, the "dispensation was granted to the Lucasian professorship in perpetuity, not just to Isaac Newton, fellow of Trinity. It was probably Barrow's last service to his protégé" (pp. 333–34). Although a diversion from the main history, this episode is significant in showing how influential the Anglican Church was in university affairs at that time.

A further example of Newton's unusual dexterity was his invention of an improved reflecting telescope. During his optical experiments he found that reflecting telescopes had produced chromatic aberration that distorted the focusing and decided to design and build a reflecting telescope that would eliminate the problem. Unlike Galileo who had hired a technician to help in constructing his telescope, Newton describes in detail how he had "cast and ground the mirror from an alloy of his own invention," along with "the tube and the mount" (p. 233). Only about six inches long, it still magnified an object "nearly forty times in diameter" (there is a Royal Society's drawing of it on p. 235). When informed of his invention and to commend his achievement he was elected to The Royal Society on January 11, 1672, the beginning of many honors to be bestowed on him.

Dr. Edmond Halley (of Halley's comet fame) conveyed the impression that it had been *his* visit to Newton in 1684 that renewed Newton's interest in celestial mechanics and instigated his writing of what is generally considered the greatest scientific work ever written, the *Principia Mathematica*. But as argued by Westfall and as indicated in Newton's account of the visit, he was already engaged in research on celestial mechanics that was a prologue to writing his book. During Halley's visit, as related to Abraham DeMoire by Newton,

> after they had been some time together, the D^r asked him what he thought the Curve would be that would be described by the Planets supposing the force of attraction towards the sun to be reciprocal to the square of their distance from it. Sr Isaac replied . . . that it would be an Ellipsis, the Doctor struck with joy & amazement asked him how he knew it, why saith he I have calculated it, whereupon Dr Halley asked . . . for his calculation. . . . Sr Isaac looked among his papers but could not find it, but he promised him to renew it, & then send it to him. . . . (p. 403)

Thus, according to Westfall, "Halley did not extract the *Principia* from a reluctant Newton. . . . The treatise *De Motu* [On Motion] which Halley received in November, bore marks testifying that the challenge was at work already" (p. 405). What surprises me about this quotation is that though seventy-five years had elapsed since Kepler had published his law in 1609, scientific knowledge was still so restricted that a person of Halley's knowledge would not have mentioned it in his original question.

Keeping his promise, Newton sent Halley a treatise of nine pages entitled *De motu corporum in gyrum* (On the Motion of Bodies in an Orbit). As Westfall states:

> Not only did it demonstrate that an elliptical orbit entails an inverse-square force to one focus, but it also sketched a demonstration of the original problem: An inverse-square force entails a conic orbit, which is an ellipse for velocities below a certain limit. Starting from postulated principles of dynamics, the treatise demonstrated Kepler's second law and third laws as well. It hinted at a general science of dynamics by further deriving the trajectory of a projectile through a resisting medium. (p. 404)

Westfall adds that "few periods have held greater consequences for the history of Western Science than the three to six months in the autumn and winter of 1684-5, when Newton created the modern science of dynamics" in *De Motu* (p. 420). When Halley read the treatise he too recognized that Newton's "celestial mechanics embodied a step forward so immense as to constitute a revolution" (p. 404).

After a series of revisions Newton gradually expanded the original nine page essay into the two volumes of the *Principia Mathematica* retaining the title *On the Motion of Bodies in Orbit* shortened to "*The Motion of Bodies*" as the subtitle of the two Books of Vol. I and "*The System of the World*" as the subtitle of Book III of Vol. II. In the spring of 1686, he gave the completed manuscript to Dr. Nathaniel Vincent, a member of the Royal Society, who brought it to the Society; and largely due to the tenacity of Dr. Halley, who oversaw the drafting of the book along with its preparation for publication, it was finally published on July 5, 1687. In the first volume Newton graciously acknowledged Halley's assistance in its final publication while Halley reciprocated with an "Ode Dedicated to Newton" that concluded with the following marvelous verse introducing volume one of the *Principia.*

> *Then ye who now on heavenly nectar fare,*
> *Come celebrate with me in song the name*
> *Of Newton, to the Muses dear; for he*
> *Unlocked the hidden treasures of Truth:*
> *So richly through his mind had Phoebus cast*
> *The radiance of his own divinity.*
> *Nearer the gods no mortal may approach.*[33]

In the *Preface to the First Edition* he presents a precise statement of his intent:

> I offer this work as the mathematical principles of philosophy, for the whole burden of philosophy seems to consist in this—from

> the phenomena of motions to investigate the forces of nature, and then from these forces to demonstrate the other phenomena; and to this end the general propositions in the first and second Books are directed. In the third Book I give an example of this in the explication of the System of the World; for by the propositions mathematically demonstrated in the former Books, in the third I derive from the celestial phenomena the forces of gravity with which bodies tend to the sun and the several planets. Then from these forces, by other propositions which are also mathematical, I deduce the motions of the planets, the comets, the moon, and the sea. (pp. xvii–xviii)

As he acknowledged, his "System of the World," although in a more concise way, recapitulates Kepler's astronomical contributions and Galileo's general program of scientific inquiry: mathematics as the language of nature, the centrality of motions, the mutual forces of gravity causing the motions of the planets, the comets, and the moon into a celestial mechanics, but he brought it to a much higher level of unification and computation. Galileo had tried to determine the velocity of light by measuring the time it took to pass from two men with lanterns some distance from each other signaling when it arrived, but its velocity was too great to be detected. Newton also rejected the instantaneous velocity of light, stating in the *Opticks* that "Light moves from the Sun to us in about seven to eight Minutes of Time, which distance is about 70000000 *English* Miles, supposing the horizontal Parallax of the Sun to be about 12"."[34] This indicates that he, like Olaf Roemer, who also measured the speed of light in 1675, were concerned to determine light's exact velocity.

Furthermore, Newton even repeats Galileo's assumption that the final answers to the mechanistic problems will depend upon understanding the causes of the motion of the insensible particles of nature, as he states in his Preface to the *Principia*:

> I wish we could derive the rest of the phenomena of Nature by the same kind of reasoning from mechanical principles, for I am induced by many reasons to suspect that they may all depend upon certain forces by which the *particles of bodies*, by some causes hither to unknown, are either mutually impelled towards one another, and cohere in regular figures, or are repelled and recede from one another. These forces being unknown, philosophers have hitherto attempted the search of nature in vain; but I hope the principles here laid down will afford some light either to this or some truer method philosophy. (p. xviii; italics added)

As he will later add in the *Opticks*: "There are therefore Agents in Nature able to make the Particles of Bodies stick together by very strong Attractions. And it is the Business of experimental philosophy to find them out" (p. 394). What a prescient statement! Perhaps influenced by Boyle's experiments to determine the constituents of water, he even suggests that by probing bodies with rays of light one might determine the size of the inherent particles. Indeed, it was due to following a similar method in the following centuries that led to the discovery of the negatively charged electron, positively charged proton, and neutrally charged neutron that would explain these attractions and repulsions, along with the atomic structure of molecules.

He begins Book I with precise definitions of concepts we now understand as 'mass,' 'inertia,' 'acceleration,' and 'centripetal force,' followed by a SCHOLIUM defining his conceptions of time, space, and motion: "Absolute, true, and mathematical time," "Absolute space," "Place as a part of space which a body takes up . . . either absolute or relative," "Absolute motion" as the "translation of a body from one absolute place into another; and relative motion, the translation from the relative place into another" (pp. 6–7). Interestingly, his motivation in adopting these absolutes was to support belief in God. As he wrote to the Reverend Richard Bentley: "When I wrote my treatise about our Systeme . . . I had an eye upon such Principles as might work wth considering men for the beleife of a Deity . . ." (Westfall, p. 441).

The evidence and arguments for these absolute conceptions seem reasonable enough at first glance, but even Newton's justification of absolute motion in relation to absolute space appears somewhat contrived. As proof of absolute motion Newton depicts a bucket half filed with water attached to a strong, lengthy rope, the surface of the water remaining flat when the rope is unwound. But when twisted and then allowed to untwist causing the bucket to rotate, the water begins to rise at the inner sides of the vessel forming a concave shape. Newton inferred from this that the "ascent of the water shows its endeavor to recede from the axis of its motion; and the true and absolute circular motion of the water, which is here directly contrary to the relative, becomes known, and may be measured by this endeavor" (*Principia,* Vol. I, p. 10).

Thus, owing to Newton, space and time were considered absolute until the Michelson-Morely experiments in 1887 and Einstein's special theory of relativity in 1905 proved that all temporal and spatial measurements, such as the simultaneity of events, were relative depending on the respective positions, velocities, and strength of the gravitational forces of the measurer and what is measured. The one exception was the constant velocity of light also regarded as the ultimate limiting velocity.

Continuing our discussion of the *Principia,* the definitions in the SCHOLIUM are followed by his presentation of the AXIOMS, or LAWS OF MOTION some of which continue to be valid to this day (within certain limits) followed by six COROLLARIES and another SCHOLIUM discussing Galileo's law of free fall. This takes us to *Book One,* THE MOTION OF BODIES, consisting of several hundred pages of complex geometrical diagrams and discussions of the mathematical relations pertaining to the various kinds of celestial and terrestrial motions. Though his supporting diagrams are usually geometrical, he occasionally uses his theory of fluxions or differential calculus when discussing magnitudes approaching "vanishing limits" or zero: for example, when explaining Aristotle's problem as to how there can be instantaneous velocities that imply motion in durationless intervals.

Newton explained this with his fluxions by showing how the rate of a dependent variables, such velocity or distance, can vanish as the independent variable, such as time becomes zero as in the differential equation ds/dt, where s stands for speed, d for distance and t for time. He also used his fluxions when he demonstrates how the attractive gravitating force of the nearest massive object deflects planetary bodies from circular to elliptical orbits permitting the deduction of the exact ratios of the distances, forces, and velocities to produce Kepler's three laws.

The second book of Volume I consists of nine sections analyzing the effects of different media on the motions of bodies along with their ratios, such as the properties of the particles in fluids affecting their fluidity as had been investigated by Boyle; the effect of air on the motion of pendulums; and how oscillating bodies in general are affected by the compression of air and the density of fluids. Though only a brief account of the scope, originality, and complexity of his investigations, it should be sufficient to convey the extraordinary range and depth of his thinking.

Turning to Volume II containing Book III of the *Principia*, including the subtitled THE SYSTEM OF THE WORLD (the title that was the basis of Hooke's charge of plagiarism because it duplicated the title he had used for one of his early works and that brought on a lasting contention), Newton originally intended it to be a nonmathematical popularization of his scientific achievements, but finally presented it in his usual mathematical rigor to avoid controversy by those unable to comprehend the mathematics. As he wrote in the introductory paragraph:

> In the preceding books I have laid down the principles of philosophy; principles not philosophical but mathematical: such, namely, as we may build our reasoning upon in philosophical

> inquiries. . . . It remains that, from the same principles, I now demonstrate the frame of the System of the World. Upon this subject I had, indeed, composed the third Book in a popular method, that it might be read by many; but afterwards, considering that such as had not sufficiently entered into the principles could not easily discern the strength of the consequences, nor lay aside the prejudices to which they had been many years accustomed, therefore, to prevent the disputes which might be raised upon such accounts, I chose to reduce the substance of this Book into the form of Propositions (in the mathematical way), which should be read by those only who had first made themselves masters of the principles in the preceding Books. . . .[35]

In insisting that his principles of philosophy be mathematical and not merely philosophical he was distinguishing himself from Aristotle and Descartes, but following Kepler, Galileo, Huygens, Hooke, and the future tradition of science. This is followed by his list of the four Rules of Reasoning in Philosophy which again are nearly identical to those stated previously by Galileo.

> Rule I: "*We are to admit no more causes of natural things than such as are both true and sufficient to explain their appearance.*" . . . Rule II: "*Therefore to the same natural effect we must, as far as possible, assign the same causes.*" . . . Rule III: "*The qualities of bodies, which admit neither intensification nor remission of degrees, and which are found to belong to all bodies within the reach of our experiments, are to be esteemed the universal qualities of all bodies whatsoever.*" . . . Rule IV: "*In experimental philosophy we are to look upon propositions inferred by general induction from phenomena as accurately or very nearly true, notwithstanding any contrary hypotheses that may be imagined, till such time as other phenomena occur, by which they may either be made more accurate, or liable to exceptions.*" (pp. 398–440)

However, he was not always consistent in following his own strict rules of reasoning. At the end of the General Scholium in Book III, he states that

> hitherto I have not been able to discover the cause of those properties of gravity from phenomena, and I frame no hypotheses; for whatever is not deduced from the phenomena is to be called an hypothesis; and hypotheses, whether metaphysical or physical, whether of occult qualities or mechanical, have no place in experimental philosophy. (p. 547)

Yet in the following paragraph he introduces the hypothesis of a "subtle spirit" pervading all bodies to explain gravity, motion, and all kinds of interactions in contradiction to his previous objections.

> And now we might add something concerning a certain most subtle spirit which pervades and lies hid in all gross bodies; by the force and action of which spirit the particles of bodies attract one another at near distances, and cohere, if contiguous; and electric bodies operate to greater distances, as well repelling as attracting the neighboring corpuscles; and light is emitted, reflected, refracted, inflected, and heats bodies; and all sensation is excited, and the members of animal bodies move at the command of the will, namely by the vibrations of this spirit, mutually propagated along the solid filaments of the nerves, from the outward organs of sense to the brain, and from the brain to the muscles. But these are things that cannot be explained in a few words, nor are we furnished with the sufficiency of experiments which is required to an accurate determination and demonstration of the laws by which this electric and elastic spirit operates. (p. 547)

As in Volume I, the rest of the book contains numerous geometrical diagrams, charts, and explanations presenting and supporting his "System of the World," which is too long and complex to be summarized here. And though I have tried to bring out the striking similarities between Galileo's contributions and those of Newton's, as I said previously, my intent is not to minimize the extraordinary achievements of Newton that guided scientific research during the following two centuries and whose formula F = ma is still used for most ordinary calculations in our familiar world and that stands with Einstein's $E = mc^2$ as the two most famous scientific formulas.

As an indication of the intense esteem the book aroused, the French mathematician of international repute, Marquis de l'Hôpital, after he had been shown a copy of Newton's *Principia,*

> cried out with admiration Good god what a fund of knowledge there is in that book? he then asked the D^r every particular about Sr I. even to the colour of his hair [asked] does he eat & drink & sleep. [I]s he like other men? & was surprised when the D^r told him he conversed cheerfully with his friends assumed nothing & put himself upon a level with all mankind.[36]

Before concluding the discussion of Newton something should be said of his later life and the publication of his final work, the *Opticks.* There were two incidents in this period that especially reveal his courageous character and integrity. The first involves the attempt by King James II, a Catholic who ascended the throne in 1685, to replace the Anglican religion with Catholicism. Hoping to accomplish this by enabling Catholics to acquire positions of authority at the universities, which was then prevented by their having to take "the oath of supremacy, in effect an oath to uphold the established Anglican religion," he decided to eliminate this obstacle by using the traditional "letter mandate" to confer higher degrees on Catholics thereby exempting them from taking the oath (p. 474).

The situation came to a climax when the King proposed Alban Francis, a Benedictine Monk, to the degree of Masters of Arts at Cambridge. When John Peachell, the Vice Chancellor, decided to resist, Newton drafted a supporting letter urging "'an honest Courage' which would 'save ye University'" (p. 475). The King on receiving the letter summoned Peachell, along with a faculty delegation to which Newton was elected as well as eight others, to appear before the Court of Ecclesiastical Commission headed by Lord Jeffreys. In a compromise the King proposed that Father Francis could be awarded the degree with the understanding that this would not be considered a precedent. Strongly objecting, Newton persuaded the delegation that this would be a dishonorable capitulation that could set a precedent.

However, when Peachell and the faculty delegation met with Lord Jeffreys and the Commission, Lord Jeffreys so intimidated Peachell that the latter failed to present a strong case for the delegation's objections and as a result resigned from the university. Thus it fell to the delegation to defend the objection with Newton forcefully advocating that they should not concede, drafting five letters preparatory to the final written response, including in one that a "mixture of Papist & Protestants in ye same University can neither subsist happily nor long together" which, however, was not included (p. 479).

Not knowing whether they would face the same fate as poor Peachell, the delegation met with Lord Jeffreys and the Commission, but this time it was the latter who yielded but warned that in the future the King's commands must be obeyed. Fortunately, the threat proved futile because eighteen months later James II was deposed by William of Orange and fled to France. But the fortitude and wise council that Newton had shown during this very threatening period did not go unnoticed or unrewarded, for when it came time for two delegates to be elected to represent the university at the convention to ratify the Glorious Revolution that deposed James II, Newton was elected as one. Then an act of Parliament led to his being one of the regular commissioners "to oversee the collection in Cambridge of aids voted to the government" (p. 480), a lucrative appointment commending his new standing.

Not only did it increase his income, but it changed his life by requiring him to move to London for a year when the convention was reconvened as Parliament. Owing to his move to London he met Christiaan Huygens and the philosopher John Locke with whom he formed a close friendship. Reading Locke's *An Essay Concerning Human Understanding* one can find many indications of what must have been their mutual influence due to their preferences for a more empirical conception of knowledge, in contrast to Descartes's rationalistic philosophy which they opposed.

As his final scholarly achievement, Newton decided to present the results of his earlier optical experiments in book form. When completed in 1694 and shown to his friend David Gregory, the latter was so impressed that he declared it "would rival the *Principia*." The Royal Society was eager to publish it but was detained owing to Newton's reluctance to have it published while Hooke was still president of the society; but when Hooke died in March 1703 and Newton was elected president the following November, he agreed to its publication by the Royal Society in 1704.

Though it did not rival the *Principia*, it was more accessible and widely read because it made less mathematical demands on the reader, had fewer geometrical diagrams, and was originally published in English rather than Latin. Nonetheless, because the questions it raised, especially the thirty-one Queries in Book III at the end of the book, were so original and far reaching that they generated much of the experimental research during the ensuing eighteenth century. As I. Bernard Cohen, professor of the History of Science at Harvard University, states, contrasting the *Opticks* with the *Principia* in his outstanding book *Franklin and Newton*:

> Not primarily in the *Principia*, then, but in the *Opticks* could the eighteenth-century experimentalists find Newton's methods for studying the properties or behavior of bodies that are due to their special composition. Hence, we need not be surprised to find that in the age of Newton—which the eighteenth century certainly was!—the experimental natural philosophers should be drawn to the *Opticks* rather than to the *Principia*. Further-

> more, the *Opticks* was more than an account of mere optical phenomena, but contained an atomic theory of matter, ideas about electricity and magnetism, heat, fluidity, volatility, sensation, chemistry, and so on, and a theory (or hypothesis) of the actual cause of gravitation.[37]

The *Opticks* consists of three Books plus the thirty-one Queries.[38] As the first three books contain a recapitulation of his optical experiments performed thirty years earlier along with some of his theories presented in the *Principia*, I shall confine my discussion to the Queries, especially as they represent Newton's genius in forecasting the scientific research of the future according to his specified methodology.

For example, in Queries 6 and 8 he remarkably anticipates the early twentieth century investigations and explanations of black-body radiation by Max Planck in 1900 and Einstein's explanation of the photoelectric effect in 1905 in stating that the reflection of light from black bodies is due to the increased intensity of the internal vibrations of the heated particles. Having indicated in Query 6 that "Black bodies conceive heat more easily from light than those of other colours," in Query 8 he asks "Do not all fix'd Bodies, when heated beyond a certain degree, emit Light and shine; and is not this Emission perform'd by *the vibrating motion of their parts*?" (p. 340; italics added). The similarity to Einstein's explanation is especially apparent since Newton's adoption of the corpuscular over the wave theory of light would conform to Einstein's interpretation in terms of discrete units of energy, called photons, rather than waves. But what was most striking was his attribution of the increased heat of the black body and ejection of the light to the intensified movement of the internal particles!

In Query 12 he presents his conclusions based on what appear to be his own experiments on the physiological nature of vision. As he states: "Do not the Rays of Light in falling upon the bottom of the Eye excite Vibrations in the *Tunica Retina* [a retinal membrane]? Which vibrations, being propagated along the solid Fibres of the optick Nerves into the Brain, cause the Sense of seeing" (p. 345; brackets added). He goes on to describe how the various magnitudes of the vibrations of the different rays of light produce in the brain the different colors, though he was mistaken in thinking that each of the optic nerves terminate in the same hemisphere of the brain as the location of the eyes, rather than crossing over to the opposite hemisphere.

In Query 28 he rejects metaphysical explanations as "feigning hypotheses," affirming that "the main Business of natural Philosophy is to argue from Phænomena without feigning Hypotheses, and to deduce Causes from Effects, till we come to the very first Cause" which "certainly is not mechanical; and not only to unfold the mechanism of the world, but to resolve these and such like Questions" (p. 369). As an indication of how difficult it is for someone even as brilliant as Newton to break completely with tradition, although having insisted on his opposition to metaphysical explanations and "feigned hypotheses," he again resorts to a transcendental explanation: "an exceedingly rare Æthereal Medium" lighter than air pervading the universe, along with God apparently as "the very first cause," to account for all the inexplicable interactions producing natural phenomena. However, now rejecting his earlier interpretation of the ether as "spiritual," he refers to it as an ethereal medium, though admitting that "I do not know what this *Æther* is" (p. 352). Despite the glaring inconsistency, the theory of a luminiferous ether filling all unoccupied space to explain the transmission of radiation such as light was generally accepted until disproved by Einstein in the early twentieth century.

In Query 31, Book III, Part I, he reaffirms the theory of particle physics consisting of the interactions of minute particles due to attractive and repulsive forces.

> Have not the small Particles of Bodies certain Powers, Virtues, or Forces, by which they act at a distance, not only upon the Rays

> of light for reflecting, refracting, and inflecting them, but also upon one another for producing a great part of the Phænomena of Nature? For it's well known, that Bodies act one upon another by the Attractions of Gravity, Magnetism, and Electricity; and these Instances shew the Tenor and Course of nature, and make it not improbable but that there may be more attractive Powers than these. (pp. 375–76)

Except for Kepler and Galileo, he was almost alone in forecasting the role of forces such as gravity, magnetism, and electricity, since most natural philosophers at the time were opposed to such so called "occult forces" acting at a distance. But this ironically was mainly due to his lifelong interest in and study of the pseudoscience of alchemy, having acquired an extensive library on the subject and secretly obtained many instruments for performing alchemical experiments. As quoted by Westfall:

> For alchemy . . . is not of that kind wch tendeth to vanity & deceit but rather to profit & to edification inducing first ye knowledge of God & secondly ye way to find out true medicines in ye creatures . . . .so y^{t} ye scope is to glorify God in his wonderful works, to teach man how to live well, & to be charitably affected helping o^{r} neighbors. (p. 298)

Not only does he condone alchemy as a sound method of inquiry, he also attributes the origin of the mechanistic worldview to an "Intelligent Agent." Returning to the *Opticks*, he states:

> Now by the help of these Principles, all material Things seem to have been composed of the hard and solid Particles abovemention'd, variously associated in the first Creation by the Counsel of an Intelligent Agent. For it became him who created them to set them in order. And if he did so, it's unphilosophical to seek for any other Origin of the World, or to pretend that it might arise out of a Chaos by the mere Laws of Nature; though

> being once form'd, it may continue by the those Laws for many Ages. (p. 402)

Rejecting Anaximander's theory that the universe arose from an original Chaos, the last statement is an affirmation of deism, the belief that though God created the universe it could then operate according to the laws he used in creating it. How Newton could combine his rigorous conception of scientific inquiry and the corpuscular-mechanistic worldview with an ardent belief in alchemy and in God as the original creator is bewildering. Though it could be argued that deducing an "Intelligent Agent" as cause of the rational laws of nature was not incoherent, a precursor of the "argument from Intelligent Design," it does not provide much of an explanation, resembling more a "feigned hypothesis." Moreover, since no particular law or causal explanation has ever been deduced from this conception nor could be subjected to any experimental test, it is arbitrary to declare "that it is unphilosophical to seek for any other Origin of the World" or that the corpuscular-mechanistic state could not have been the original state as is now believed by most astrophysicists.

Though necessarily limited, I hope this was sufficient to convey an authentic portrayal of Newton's genius as well as his idiosyncrasies. Returning to the account of his final years, after publishing the *Opticks* he decided to leave the scholarly life he had preferred for thirty-five years at Trinity College to move to London permanently. However, even his recently increased income was not sufficient to support living in London, so he had to seek a more profitable employment. Encountering the usual initial disappointments, with the aide of his friend Charles Montague he succeeded in obtaining the lucrative position of Warden of the Mint in March 1696. Three and a half years later, thanks to his skillful management, he was promoted to Master of the Mint, a position he held until his death. This settled his financial problems because he received the considerable sum of "a set profit on every pound weight troy that was coined," along with his salary of 500 pounds per annum (Westfall, p. 604). Thus he resigned his fellowship and the Lucasian Chair at Cambridge the following year.

Surprisingly, however, despite his renouncement of his academic career for a more active social and political life, in 1696 when he became Warden of the Mint he showed that he had not lost all his interest in nor remarkable mathematical skills when he replied to two challenge problems sent to him by Johann Bernoulli, a prominent European mathematician, that had been published in John Wallis's *Acta eruditorum* (Acts/Reports of the Scholars). Soon after receiving his copy of the problems Newton anonymously returned the solutions to Bernoulli who easily identified their author due to his method of solving them. Only two other renown mathematicians, Leibniz and the Marquis de l'Hôpital, submitted solutions indicating that Newton had not lost his acute mathematical ability despite his age and change of vocation.

As indicated previously, the year Hooke died (1703) Newton was elected president of the Royal Society and as usual served with distinction enlarging the membership and attendance, improving the quality of the lectures, and increasing the financial support. Aided by Secretary Hans Sloane, he was able to persuade the members to purchase the former house of Dr. Edward Browne on Crane Court, London, in 1710, for its permanent residence which was paid for in less than six years. Later the society was moved to the attractive row of white buildings on 6–9 Carlton House, where I was shown the original manuscript of Newton's *Principia* by Keith Moore.

Following his election to the presidency of the Royal Society, Newton would live for another quarter century less one year, generally in good physical and mental health until five years before his death, though continuing as Master of the Mint and president of the Royal Society to the end. During these later years, except for his solution of Bernoulli's challenge problems, he completely replaced his previous dedication to natural philosophy and mathematics by religious studies. But he is renown not for his alchemical or religious interests, but for his outstanding contributions to physics, astronomy, and mathematics.

Yet for all his extraordinary achievements, unlike Francis Bacon and Descartes he always retained his modesty. He is admired for his self-deprecating assertions, such as "If I have seen further it is by standing on ye shoulders of giants" (Westfall, p. 274); "I don't know what I may seem to the world, but, as to myself, I seem to have been only like a boy playing on the sea shore, and diverting myself in now and then finding a smoother pebble or a prettier shell than ordinary, whilst the great ocean of truth lay undisclosed before me" (Westfall, p. 863); and, finally, as death approached, "He said when he died 'he should have the comfort of leaving Philosophy less mischievous' than he found it" (Westfall, p. 805).

He died on March 20, 1727, after a brief illness at age eighty-five. He is prominently interred in the nave of London's prestigious Westminster Abbey. A monument to him bears the fitting inscription: "Let Mortals rejoice that there has existed such and so great an Ornament to the Human Race." Thus ends the chronicle of one who had the intelligence, courage, creativity, and perseverance to consolidate a methodology and worldview that, though initiated by Kepler, Galileo, and others, led to what was one of the greatest transformations in our conception of reality and way of life that established scientific inquiry as the only known sound method of our investigation of the world.

Yet there was and still is considerable opposition to these scientific arguments supporting a more naturalistic worldview. As Carl Sagan has pointed out in his extraordinarily informative book *The Demon-Haunted World: Science as a Candle in the Dark*, the question whether the world could be best explained by an empirical-rational approach utilizing the language of mathematics has been contested by the fact that human beings historically have been more influenced by nightmares, fantasies, hallucinations, and visitations by demons, monsters, devils, spirits, angels, and gods that helped form primitive myths, sacred documents, and religions, along with witch hunts and alleged heretical immolations that influenced by technology, have been transmuted into more familiar extraterrestrial aliens, flying saucers, and UFOs.

As an example, Sagan quotes on his page 129 W. Gary Crampton's later statement in the May 23, 1994, issue of *Christian News* that, despite the enormous advances in science in the twentieth century, the Bible still remains the source of ultimate truth.

> The Bible, either explicitly or implicitly, speaks to every area of life; it never leaves us without an answer. The Bible nowhere explicitly affirms or negates intelligent extraterrestrial life. Implicitly, however, Scripture does deny the existence of such beings, thus also negating the possibility of flying saucers. . . . Scripture views the earth as the center of the universe . . . According to Peter, a "planet hopping" Savior is out of the question. Here is an answer to intelligent life on other planets. If there were such, who would redeem them? Certainly not Christ. . . . *Experiences which are out of line with the teachings of Scripture must always be renounced as fallacious. The Bible has a monopoly on the truth.* (italics added)

It is difficult to see how any intelligent person, given the advances in science and what we know about the world today, could still believe that the Bible "has a monopoly on truth."

Chapter IV

THE EIGHTEENTH AND NINETEENTH CENTURIES' ADVANCES, INCLUDING INQUIRIES IN MAGNETISM AND ELECTRICITY

While I. Bernard Cohen's claim that the eighteenth century was the "age of Newton" has considerable justification, it overlooks other social and cultural changes that were also essential in shaping the important scientific transformations that led up to the enlightenment of the eighteenth century. We have already witnessed the crucial role that previous technological innovations, such as the invention of the telescope, microscope, astrolabe, etc., along with such mechanical models as Kepler's clockwork universe and Galileo's microscopic particles culminating in Newton's corpuscular-mechanistic worldview played in creating modern classical science. In addition, the earlier Renaissance transition in Europe that rejected the medieval outlook and religious authority as bastions of truth also contributed to these scientific developments that evoked a more empirical, rationalistic outlook, yet these changes still were confined largely to contribution of scholars trained at the major universities.

We shall find that other institutions of change in the eighteenth century, the Age of Enlightenment, which were broader in scope and diversified in nature and not confined to the traditional educational regimes of the seventeenth century, also played a major role. So along with the significant influence of Newton there were other cultural changes, such as the impact of the industrial revolution that introduced new technologies like spinning and weaving machines and steam engines that replaced home guilds by factories and transformed agrarian societies into crowded cities with their slums and unsanitary conditions, along with amenities.

The Industrial Revolution also changed the economic system by introducing mass production and an entrepreneurial and capitalistic economy in the industrial cities, along with a new middle class to replace the traditional aristocratic-peasant division of society. Furthermore, there was a marked change in the religious backgrounds of those engaged in the new investigations from those who were Anglicans and had graduated from the usual prestigious preparatory schools such as Eton and universities like Oxford and Cambridge, to those who were often self-taught and learned from their particular trade, along with being educated in what were called "Nonconformist" educational institutions. As I wrote previously:

> Underlying this change in background of the new generation of natural philosophers was the shift of the centers of experimental research from Oxbridge to provincial manufacturing towns like Birmingham and Manchester, the latter with its burgeoning textile industry becoming the leading manufacturing and trade center of the world. The advent of the industrial revolution with its large cotton and woolen mills, as well as expanding industries such as coal mining, ironworks, and steel foundries, not only attracted huge numbers of workers and created a new middle class of wealthy families, it rewarded manual skills, engineering inventiveness, and entrepreneurial shrewdness—the type of practical intelligence characteristic of the new breed of natural philosophers.[39]

As examples, two of the outstanding English chemists of the period were the Unitarian Minister Joseph Priestley who was the son of a cloth shearer and John Dalton a Quaker whose family income came from cottage weaving, along with the American Benjamin Franklin who, despite leaving school at age ten to help his father who was a tallow chandler and soap maker, became a world famous writer, printer, and statesman, as well as attaining international renown for his electrical experiments.

Thus the new experimental inquiries and industrial ventures were largely made by nonconformists like Quakers, Presbyterians, and Unitarians, as well as the sons of craftsmen whose interests or endeavors were not restricted to the traditional classical studies, but motivated by utilitarian interests and needs. Unable to attend the traditional grammar schools and universities because of their lower family backgrounds and inferior religious affiliations, "Dissenting Academies" were created to meet their different educational requirements. In Birmingham such eminent industrialists, inventors, and natural philosophers as Josiah Wedgwood, Joseph Black, Mathew Boulton, John Wilkinson, Erasmus Darwin (grandfather of Charles Darwin), and Joseph Priestley founded the Warrington Academy in 1757, the most famous of the dissenting academies, and the prestigious Lunar Society about 1775.

In the early eighteenth century, owing to Newton's discussion of electricity and magnetism in his Queries, these subjects attracted some scientific interest. Newton had included them among the important forces of nature that emanated in space, noting that they attracted small particles at a distance, a fact considered by the ancient Greeks, though still unable to explain how they were produced. The Greeks had discovered that rubbed amber, called "electron" in Greek (from which the term "electric" is derived), along with "Heraclean stones" (later called "lodestones" or "magnets") had these attracting powers. Yet at the time of Newton, because Descartes and most natural philosophers disavowed "occult powers" acting at a distance, believing that the transmission of forces required some material medium or direct physical contact, it was primarily William Gilbert's book, *De Magnete* (On the Magnet) published forty-two years before Newton was born, that brought attention to magnetism and electricity owing to their mysterious forces acting at a distance.

Gilbert was born in 1540 in Colchester County, Essex England. His background is somewhat uncertain, but he is said to have attended the Grammar School of Colchester and enrolled in St. John's College, Cambridge where he apparently earned BA, MA, and MD degrees. After those studies he traveled on the continent where he may have acquired the Doctor of Physic Degree. In the biographical memoir to his translation of the *De Magnete,* P. Fleury Mottelay states:

> In 1573, he was elected a Fellow of the Royal College of Physicians, and filled therein many important offices, becoming in turn [omitting the dates] Censor, Treasurer, and President. His skill had already attracted the attention of queen Elizabeth, by whom he was appointed her physician-in-ordinary, and who showed him many marks of her favor . . . ."[40]

As an indication of how he was regarded at the time, the English historian Henry Hallam wrote that he was

> the first in which England produced a remarkable work in Physical Science: but this was one sufficient to raise a lasting reputation for its author. Gilbert, a physician, in his Latin treatise on the Magnet not only collected all the knowledge which others had possessed on the subject, but became at once the father of experimental philosophy in this island, and, by a singular felicity and acuteness of genius, the founder of theories which have been revived after a lapse of ages, and are almost universally received into the creed of science. . . . (p. xii)

When *De Magnete* was published in 1600 it was highly acclaimed because of its exhaustive summary of the previous investigations into magnetism, as well as Gilbert's own contributions. Galileo declared Gilbert "great to a degree that is enviable"; Dr. William Whewell, a prominent philosopher of science in the nineteenth century, observed that "Gilbert's work contains all the fundamental facts of the science, so fully examined . . . that even at this day we have little to add to them"; and Dr. Thomas Thomson said "that *De Magnete* is one of the finest examples of inductive philosophy that has ever been presented to the world" (pp. xii, xiii).

At the time Gilbert's *De Magnete* was thought to contain all that was known about the magnet, including his own discoveries. He detected a magnetic force in numerous substances that he called "electrics" and according to historian of science Sir David Brewster, applied "the term *magnetic* to all bodies which are acted upon by loadstones and magnets, in the same manner as they act upon each other, and he finds that such bodies contain iron in some state or other" (p. xv). In Gilbert's study of electricity he is credited with introducing for the first time such terms as "electric force," "electric emanation," and "electric attraction" (p. xv). He noted its considerable resemblance to magnetism, yet distinguished between the two based on discovered differences.

He regarded the earth to be a vast magnet attributing to its extremities north and south magnetic poles, "calling *south pole* the extremity that pointed toward the north, and *north pole* the extremity pointing toward the south" (p. xv). Noticing the inclination and declination of the magnetic needle, he inferred that at the opposite poles the needle would extend vertically, at the equator horizontally, and in between in intermediate directions. This led to the European perfection of the compass and the designation of latitudes and longitudes by the dips in the needle that vastly improved navigation. He created a globe named '*terrella*' or little earth that could be used in his experiments to represent the poles and latitudes and longitudes of the earth. He also invented what is called a *versorium,* known as an electroscope, a device consisting of a rotating iron needle positioned on a movable point so that it could rotate freely enabling him to measure the intensity of electrical discharges (p. 79).

Ironically, despite this acknowledged acclaim at the time and that three Latin editions soon appeared with elaborate title pages, it neither commanded much attention in England nor stimulated continuing research until about three centuries after its publication. As Mottelay writes at the very beginning of his translation, "I FIRST entered upon the translation of this, the earliest known published work treating of both magnetism and electricity, in the beginning of 1889" (p. v). Robert Boyle did publish a brief essay on electricity in 1675 making extensive use of the recently invented vacuum pump and insisting that Gilbert's "electric effluvium" was composed of minute particles possessing the attractive property, nothing being known yet about its repulsive power.

Thus it was not only Gilbert's book but the research instigated by Boyle and particularly Newton who, as President of the Royal Society, had incited by his Queries the succeeding tradition of electrical research. One of his first acts on becoming president of the Royal Society was to appoint Francis Hauksbee as Curator of Experiments, despite his lack of a formal education, based on his experimental papers on electricity published in the Society's *Philosophical Transactions,* thus illustrating my previous claim regarding the backgrounds of the new researchers. As stated by the Duane and Duane H. D. Roller in *The Development of the Concept of Electric Charge:*

> One may guess that his skill at constructing instruments and his unusual genius for experimentation were what brought him into association with the members of the Royal Society . . . to prepare experiments and was paid for doing so. The facilities and associations afforded Hauksbee by the Society must have been a factor in helping him become "the most active experimentalist of his day."[41]

Indeed, at Newton's very first meeting presiding over the Royal Society Hauksbee demonstrated his perfected instrument crucial for his electrical experiments, the air pump used to created a vacuum. At the time there was considerable interest in streaks of light, called barometric light, that appeared above the mercury inside a barometric tube when shaken. Hauksbee discovered that the flickers of light were only produced when the drops of mercury slid down the glass, never when stationary, and only begun when about half the air was removed, increasing in brightness with the rarification of the air but discontinued entirely when all the air was withdrawn. Apparently surmising that there might be some similarity between these illuminations and electricity, he devised an experiment in which in an evacuated chamber he caused amber to be rubbed against wool that produced similar results, suggesting that the barometric light might be electrical.

Next he evacuated a sealed glass globe and found that in rotating it by rubbing it with his hands flashes of light again appeared in the evacuated interior. He then found that a nonevacuated globe when rotated with his hands produced electrification on the surface of the globe, not in the interior. He next discovered that this "charged" glass globe could attract or repel a brass leaf and that when held to his face produced a sensation like an "electric wind." While detecting that an electric globe electrified a neutral one, he did not realize that this was an example of induced electrification believing instead, based on his effluvial theory, that it was due to the attrition of the effluvial material.

In another ingenious experiment showing how important experimentation was becoming in trying to solve technical problems, he surrounded a central globe with a semicircular wire from which he hung a line of threads above the surface of the globe. He found that with an *uncharged* spinning globe the movement of the air aligned the threads in the same direction as the air, but that a sufficiently *charged* globe attracted the threads in the direction of the central globe overcoming the force of the air. Then, pointing his finger at the loose ends of the circular band of threads facing the globe, he found they were repelled, but if directed to the top loop of the threads they were attracted, though he did not recognize the difference between attractive and repulsive forces, thus showing how difficult it is to break with tradition. In summary, while

> Hauksbee's experiments linked barometric light with electric effects, introduced the triboelectric generator, demonstrated the occurrence of electrical influence, and provided evidence of electrical repulsion as well as attraction, his attempts at explaining the phenomena in terms of a material effluvium [perhaps a precursor of Newton's Æthereal medium] were week, illustrating the importance for scientific progress of theorizing as well as ingenious experimentation.[42] (brackets added)

Little is known about the early background of the next contributor, Stephen Gray (1666–1736), other than that he lived as a "poor brethren" in the charterhouse and therefore also was unable to acquire a formal education. To convey an idea of life at the time, the Charterhouse was founded to provide schooling for boys who were "gentlemen by descent and in poverty" and a living for poor brethren who were preferably "soldiers that had borne arms by sea or land, merchants decayed by piracy or shipwreck, or servants in household to the King or Queen's Majesty" (p. 571). As with Hauksbee, what we know about Gray is limited to his communications with the Royal Society, although how he became involved in electrical experiments is unknown.

Gray's initial published paper on electricity appeared in the *Philosophical Transactions* in 1720, then nine years later he divulged to Dr. John Theophilus Desaguliers (Newton's assistant) and others his discovery that the "electric virtue" of a rubbed glass tube could be conducted by a "packthread" over great distances, more than 650 feet, detecting what is now called "electrical conduction" (p. 330). He also found that the success of the conduction depended upon the nature of the connecting material and anticipated the distinction between "electrical *per se*" and "nonelectrical" noting that rubbing could create the former but that the latter could not be produced by rubbing but only by coming in contact with an electrified body (p. 330).

Anticipating Benjamin Franklin, he also found that when a metal rod with a pointed tip was brought to an electrified object it attracted the electricity in a smooth and silent manner, while a blunt tipped rod produced a bright flash with a sharp snap. But his most important contribution was the discovery that whatever the nature of the "electric virtue" or "electric fluid," as the electricity was then called, when created it *exists independently* of the charged source and thus constitutes a separate entity, like gravity, magnetism, and heat. This was reinforced by finding that an unelectrified object can be electrified by bringing it close to an electrified object without touching it, suggesting that the electric virtue by itself existed between the two objects. Again anticipating Franklin, he suggested that this "electric fire" resembles thunder and lightening. For his exemplary research in 1732 he was made a Fellow of the Royal Society (FRS), twelve years after his first submission to the Royal Society (p. 331).

The next contributor not only benefited from the previous electrical research of Hauksbee and Gray, but also from a better education becoming a member of the French Academy of Sciences and a Fellow of the Royal Society. His extended French name, Charles François de Cisternay du Fay has been abbreviated simply to Dufay. Having studied Gray's experiments eight months earlier, by December 1733 he was able to submit to the Royal Society a summary of his own experiments under seven headings, the first five of which were an extension of the previous research while the sixth presented the discovery of a simple but significance principle of electrification that holds true today. "This principle is that an electrified body attracts all those that are not themselves electrified, and repels them as soon as they become electrified by . . . [conduction from] the electrified body" (p. 331).

Then, due to his asking the acute question of whether the repulsion was restricted just to bodies electrified in the same manner or also applied to those that had been electrified differently, he discovered that when rubbed glass was brought in contact with an electrified resinous substance such as copal, it was not repelled as expected but attracted. Thus his seventh discovery consisted of finding that there

> are two distinct electricities, very different from each other: one of these I call *vitreous electricity*, the other, *resinous electricity*. The first is that of [rubbed] glass, rock crystal, precious stones, hair of animals, wool, and many other bodies. The second is that of [rubbed] amber, copal, gum, lac, silk, thread, paper, and a vast number of substances]." (p. 333; italics and brackets in original)

It was not then known that the type of electrification produced depends not only on the material of the electrified objects, but also on the nature of the material used in the rubbing: wool, silk, or cat's fur producing vitreous electricity while rabbit's fur creates resinous electricity.

Thus Dufay made the important discovery that the same kinds of electrics will repel each other while the opposite kinds will attract. In addition he found that neutral or unelectrified objects, if they can be electrified, can be electrified by either kind of electricity. Though he normally did not speculate about the nature of electrification, instead of accepting an effluvium surrounding the electrified objects he introduced an *atmosphère particulière*, an "electric fluid" (analogous to the release of caloric fluid or phlogiston in explaining combustion), as the source of the peculiar manifestations when electricity is transmitted from one electric object to another. He also surmised that neutral objects contain an equal amount of the different kinds of electrical fluids ("vitrious" and "resinous") and that when two objects with different amounts of electricity come in contact, the one with the greater amount will transmit its excess to the lesser one until equilibrium is reached (pp. 333–34).

In Europe the concept of electricity as a fluid produced a flurry of experiments related to its being contained and transmitted like a fluid. Though somewhat different, the concept of an electric "charge" was also introduced based on the analogy of "charging" armaments with gun powder. The term "electric charge" is still retained. A number of experiments, highly dangerous, were performed to show how the electric fluid could be contained and the quantity measured.

Another experimenter was a Pomeranian clergyman named E. G. von Kleist. In 1745 Kleist, an amateur experimenter, performed an experiment with dramatic results showing how electricity could be contained. Using a bottle containing water with a very narrow neck enclosing a nail, while continuing to hold the bottle he electrified the nail and then bringing it in contact with an unelectrified object it produced an intense spark. Still holding the bottle, he touched the nail with his other hand experiencing such a severe shock that "it stuns my arms and shoulders." If the bottle were removed from other objects it remained charged for some time, showing that it retained the electrification (pp. 334–35).

In 1746 Dutch professor Pieter van Musschenbroek, in another shocking experiment, conveyed the results to the French Academy warning others not to try it. He suspended the barrel of a gun by two long silk threads at each end. Electrifying one end of the barrel from the other he hung a brass wire extending into a glass flask, partially filled with water. "This flask I held in my right hand, while with my left I attempted to draw sparks from the gun barrel. Suddenly my right hand was struck so violently that all my body was affected as if it had been struck by lightning. . . . The arm and all the body were affected in a terrible way that I cannot describe: in a word, I thought it was all up with me . . ." (pp. 334–35). Because of the notoriety of the experiment and since Musschenbroek was a professor at the University of Leiden, it became known as the Leiden experiment and the flask, named the "Leyden jar," was used in chemistry laboratories when I was a student and is still used today.

Performing electrical experiments had become so popular by the mid eighteenth century that they were conducted by amateurs as well as natural philosophers and were described in popular magazines as well as scholarly journals. The interest was so extensive that it even reached the North American colonies, coming to the attention of Benjamin Franklin, known to many as "America's first great man of science." Despite leaving school at age ten to assist his father in his workshop, Franklin became famous as a writer, printer, diplomat, and experimental physicist. Like those in Manchester and Birmingham, he was instrumental in organizing the American Philosophical Society (the first society in the colonies for the discussion of scientific topics); helped establish the Library Company, the earliest lending library in America; and was one of the founders of the Philadelphia Academy and College that later became the University of Pennsylvania.

He was in his late thirties when, in 1743, having attended lectures in Boston by Dr. Spencer from Scotland, he became interested in electrical experiments. Receiving a present of a glass tube from Peter Collinson, a Fellow of the Royal Society, he wrote in his autobiography, "I eagerly seized the opportunity of repeating what I had seen in Boston, and, by much practice, acquir'd great readiness in performing those also which we had an account of from England, adding a number of new ones" (p. 337).

Having achieved financial independence, he was able to devote full time to his experiments and purchase any equipment needed for his research, including the newly invented Leyden jar. In subsequent communications with Collinson in which he describes his initial experiments and in turn receiving reports of electrical experiments abroad, Franklin presents in some detail an experiment on how electrical fluid, or as he called it "fire," was variously conducted among a number of persons standing on wax insulators.

Without going into the details of the experiments, I shall just relate what he and his collaborators contributed that included the significant introduction of the terms "positive" or "plus" and "negative" or "minus" electrics. As he states:

> Hence have arisen some new terms among us. We say *B* (and bodies like circumstanced) is electrized *positively; A, negatively.* Or rather, *B* is electrized *plus; A, minus.* And we daily in our experiments electrized [objects] *plus* or *minus,* as we think proper. To electrize *plus* or *minus,* no more needs to be known than this: that the parts of the [glass] tube or sphere which are rubbed do, in the instant of the friction, attract the electrical fire, and therefore take it from the thing rubbing; the same parts immediately, as the friction upon them ceases, are disposed to give the fire they have received to any body that has less. (pp. 340–41; brackets in original)

Despite the fact that Franklin and his associates benefited from the research of others in England and Europe, that in less than four years they were able to formulate a conceptual framework that generally accounted for the experimental results was a remarkable achievement, especially its quantifiability. This was illustrated in their explanation of the function of the Leyden jar. Because on their theory the total electrification was conserved, they showed experimentally that when the inner coating of the jar was positively electrified the outer coating was equally charged negatively, the flow always going from the greater to the lesser amount of electrification, but if connected by a wire equilibrium was instantly established.

Then, in seeking a more fundamental explanation, in a paper entitled "Opinions and Conjectures Concerning the Properties and Effects of the Electric Matter, arising from Experiments and Observations made at Philadelphia, 1749," he attempted to explain the "electric matter" or "fluid" as consisting of very subtle particles since it penetrated all substances including hard metals. Moreover, if these particles are pliable and repel each other, then the repelling effect can be attributed to them. But while normally repellent, if they come in contact with a neutral object they will be distributed uniformly to maintain the neutrality, while if a conductor loses particles by their being attracted by another object, the remaining particles will attract additional particles to maintain equilibrium.

In answer to a criticism as to how objects can acquire an excess of electric matter if they can retain only a quantity equal to their own particles, Franklin claimed that "in common matter there is as much electric matter as it can contain; therefore, if more be added it can not enter the body but collects on its surface to form an 'electric atmosphere,' in which case the body 'is said to be electrified.'"[43] Yet there was a remaining objection. While it was obvious why two positively electrified objects repel each other, when it was discovered that two negatively charged objects also repel, there was no immediate explanation. Why should two bodies possessing less electricity resist sharing some electricity?

As usual when one encounters an anomaly in a theoretical explanation something either has to give or be added. In this case it was German natural philosopher Franz V. T. Æpinus who introduced a resolving assumption.

> The revolutionary idea of Æpinus was that in solids, liquids, and gases the particles of what Franklin called "common matter" repel one another just like the particles of the electric fluid in Franklin's theory. Æpinus's revision introduced a complete duality. The particles of common matter and of electric matter each have the property of repelling particles of their own kind while each kind of particle has the additional property of attracting particles of the other kind. (p. 343)

Attributing additional electrical charges to the natural particles of a body came to be known as the "two fluid system" analogous to Dufay's earlier hypothesis of two electric fluids, one vitreous and the other resinous.

However, as usually occurs with scientific explanations, while Æpinus's resolution explained the anomaly of negatively charged bodies repelling each other despite having fewer electrically charged particles, this explanation raised a further problem, as Æpinus realized. If the particles of common matter also repel each other this conflicts with Newton's universal law of gravitation that all material objects exert a gravitational attractive force on each other. How could the repulsive force of the negatively charged common particles generate the attractive gravitational forces? Æpinus proposed a counter-explanation to no avail; the resolution was beyond an explanation at the time that would have to await the discovery in atomic physics of variously charged particles.

But despite the theoretical impasse there was sufficient truth in Franklin's conception of electricity that he was able to draw practical consequences from it that enhanced his international acclaim. Although he was not the first to suggest that there was an affinity between electricity and lightning, he was the first to establish their identity. In an entry in his "experimental notebook" he indicated that there were twelve ways the "Electric fluid agrees with lightning:"

> (1) giving light; (2) color of the light; (3) crooked direction; (4) swift motion; (5) being conducted by metals; (6) crack or noise in exploding; (7) subsisting in water or ice; (8) rending bodies it passes through; (9) destroying animals [he has killed fowls by the discharge of several Leyden jars connected together]; (10) melting metals; (11) firing inflammable substances; (12) sulfurous smell.[44]

In the essay on "Opinions and Conjectures Concerning the Properties and Effects of the Electric Matter" mentioned earlier, he had indicated that pointed objects attract an electrical force at a greater distance and with greater ease than a blunt object. He also expressed his belief that clouds were electrified as seen in bolts of lightning. But never satisfied with just conjectures, these combined documents led him to devise means of testing whether lightning was truly electrical and that clouds too were electrified, along with inventing ways of avoiding being struck by them.

Thus with the help of his son he undertook his famous kite experiment to prove that lightning was indeed a form of electricity. After attaching a wire as the detector to the front of a kite, he then tied to it a long kemp cord that reached the ground on the end of which was fastened a key and a silk ribbon for insulation. At the outbreak of a storm they raised the kite and ran into a shed after allowing the cord to become wet to increase its conductivity. Holding the kite by the dry silk ribbon so as not to be electrified, as expected a bolt of lightning from a passing cloud struck the wire detector and was transmitted through the cord to the key where it was then collected into a Leyden jar as proof of its electrical nature.

In another variation of the experiment, attaching a pointed metal rod to the peak of his roof he hung from it a long wire descending from the side of the house to a metal frame holding two iron bells with metal clappers. As before, when lightning struck the electrical discharge it was transmitted to the rod down the wire to the iron bells which produced a clanging sound. Owing to these experiments, lightning rods were installed on the tops of buildings and church spires to deflect lightning from striking them and causing a fire. Although others had thought of the possibility of such experiments and protective devises Franklin was unique in actualizing them.

In conclusion, as stated by Duane and Duane H. D. Roller in the book cited previously:

> By 1757 the public demands on Franklin's time had become so great that he ceased completely the experimentation that had already earned him the reputation of the foremost electrical scientist of his day.
>
> By this time he had received the Copley Gold Medal, which is the highest distinction that the Royal Society can bestow, and had also been elected a Fellow of the Society. In 1773, the French Academy of Sciences made him a "foreign associate," an unusual honor and one that was not to be accorded to another American scientist until a century later. (p. 607)

The growing confidence in the progress of science due to acquired scientific explanations confirmable by experimental evidence and expressed in the language of mathematics, as proposed by Galileo and Newton, having been reinforced by the electrical investigations, especially Franklin's quantification of electrical phenomena, there followed an attempt to ascertain whether one could discover electrical laws comparable to Newton's universal laws of gravitation. Based on the analogy with Newton's law that gravity is a function of mass, distance, and gravitational forces, perhaps electricity can be measured in terms of mass, distance, and *electrical* forces.

In fact the Swiss physicist Daniel Bernoulli invented an electrometer that directly measured the strength of "the electric force between two charged metal disks when they were at known distances apart," the measurements indicating that "the force varies inversely as the square of the distance between the plates," conforming to Newton's gravitational law (p. 610). In another experiment Joseph Priestley, the identifier of oxygen, also confirmed that the strength of the electric force agreed with Newton's inverse square law. Then French physicist Charles Coulomb devised an "electrical torsion balance" that proved so accurate that he could measure "with the greatest exactitude the electrical force exerted by a body, however slightly the body is charged" (p. 617), confirming that the strength of the *repulsive* force between two equally electrified bodies varies inversely with the square of the distance.

Again, according to Duane and Duane H. D. Roller, since Newton's law applies to the *attractive* force between two objects this also had to be confirmed, as Coulomb succeeded in doing with an "electric raised torsion pendulum" that he also invented (p. 620). The question then was whether the other portion of Newton's law also applied, that the gravitational force was proportional to the product of the masses (density per volume). Though the concepts of "electric fire" or "electric fluid" would not seem amenable to such a confirmation, Franklin's hypothesis of bodies being composed of two kinds of particles (a kind of matter) that exert opposite forces, negative for electric and positive for natural matter, suggested that there was the possibility of a determination if one substituted electric mass" for "gravitational mass."

Believing it was possible, Coulomb declared that the "electrical force between two electrified objects is proportional to the inverse square law of the distance between them and to the product P of their electrical masses, or $f \mu P/ d^2$" (p. 621), again conforming to Newton's law of gravitation and that became known as "Coulomb's law." As Duane and Duane H. D. Roller state:

> With this quantification of electrical science, it becomes possible to bring to bear upon its further study the entire weight of mathematical techniques. Eighteenth-century mathematics had to a very large degree developed along lines applicable to Newtonian mechanics, and with the formulation of electrical science in quantitative terms so analogous to mechanics, electricity became thoroughly amenable to mathematical treatment, with striking results in the nineteenth century. (pp. 621–22)

Turning next to the investigation of light, throughout history fire, sun, and sunlight have been of intense interest. The sun was deified as *Helios* and *Sol* respectively by the ancient Greeks and the

Romans. The Pythagoreans placed fire in the center of the cosmos calling it the "Hearth of the World" and the "Throne of Zeus." Plato in the *Republic* declared that "of all the divinities of the skies the sun is the most glorious because it not only . . . gives to the objects of vision their power of being seen, but also their nourishment and existence."[45] It was partly due to its exalted position that Copernicus and Kepler ceded to the sun its central place in the universe, though little was known then about the nature of light and its transmission.

By the time of Newton the two dominant theories of the transmission of light were Descartes's view that light as seen was the physiological effect on our senses of the "*instantaneous* pression" of the contiguous motionless particles comprising the fluid vortices of the universe while the other was the wave theory of light held by Robert Hooke, Christiaan Huygens, and others. Yet for reasons presented in our previous discussion of the *Opticks*, Newton rejected both theories based on his prismatic experiments and corpuscular theory of light. So just as Newton's Queries in the *Opticks* stimulated research into the theories of an ethereal medium, gravity, particles, magnetism, and electricity, in the eighteenth century, his Queries from 21 to 31 discussing the properties and transmission of light, especially that it consists of rays composed of corpuscles, encouraged investigations into optics and light.

Though his theory had gained ascendance by the early eighteenth century, it was challenged by Thomas Young in a paper entitled "Outlines of Experiments and Inquiries Respecting Sound and Light" published in the Royal Society's *Philosophical Transactions* in 1800. Drawing an analogy with the transmission of sound, Young rejected the particle theory of light in favor of the wave theory that depicted light, like sound, as the undulations of an underlying stationary medium.

Young presented several objections to the corpuscular theory before providing the main evidence in favor of the wave theory.

The first was that if light were composed of material particles they would be attracted by gravitational forces so that their velocities would vary with the strength of the gravitational force of the emitting body, yet light seems to travel with an invariant velocity; however, this objection does not apply to waves which, if propagated in an aetherial medium, are not affected by gravity. Second, Young believed that Newton's explanation that the light and dark rings of light, known as "Newton's rings," are caused by the partial reflection and transmission of the light particles when directed through two lenses separated by a film of air, described as "Fits of easy Reflection and easy Transmission," could more reasonably be explained by the refraction of alternating light and dark waves.

As additional support he found that when monochromatic light is projected on a screen that has a circular opening the diameter of which is larger than the wavelength of the light passing through, it produces a circular image on the screen behind it. But if the diameter of the opening is about equal to the wave length of the light then a series of alternating light and dark bands indicative of the interference of waves appear on the posterior screen. He found that the latter effect is produced also when two holes very small and close together are cut in the screen and a beam of monochromatic light strikes the screen midway between the two points. In an essay titled "On the Theory of Light and Colors," published in 1802, he described these bands as being "constructive (in phase) and destructive (out of phase) interference."[46]

Despite the nearly incontrovertible evidence, as an indication of how strong Newton's influence was at the time and how difficult it is for even some scientists to question or reject their theories, Henry Brougham described Young's paper as "destitute of every species of merit . . ." (p. 19). However, Young's conclusions did resonate in the thinking of a gifted French engineer, Augustine Jean Fresnel, who rejected the corpuscular explanation of diffraction declaring that it had been refuted experimentally. When, in defense of the corpuscular theory, the supporters maintained that the diffraction patterns in Young's experiment were produced by the edges of the circular holes deflecting the particles passing through, Fresnel tested the explanation by altering the shape and the mass of the holes and finding that it had no effect at all, the diffraction pattern depending only on the relative sizes of the apertures and the wave lengths of the monochromatic waves.

He then supplemented Young's experiments by attributing mathematical dimensions to the properties of the diffracted waves. As we now know, the properties of particles and waves are the converse of each other: particles having a discrete location in space with various shapes and sizes, possess mass and momentum along with the energy of motion, and interact by deflection with a loss of energy. In contrast, waves are defined by their lengths, frequencies, amplitudes, and intensities, are diffused in space as wave trains, and interact to reinforce if in phase or destruct if out of phase. As described by Peter Achinstein:

> [Fresnel's] account is much more sophisticated than Young's, not only because it is quantitative, but also because in determining the resulting vibration . . . Fresnel derives mathematical expressions for the amplitude of the vibration at any point behind the diffractor, and for the light intensity at that point. From these he infers the positions and intensities of the diffraction bands—inferences that were confirmed experimenally. (p. 21)

Fresnel was awarded a prize when he sent his results in a "Memoir on the Diffraction of Light" to the Paris Academy in 1819.

As optical investigations continued further evidence was discovered to support the wave theory. The initial inability of the wave theory to explain the polarization of light emerging from Iceland spar due to the assumption that light waves were transmitted longitudinally, running lengthwise like sound percussions, was surmounted when they were discovered to be produced by transversal vibrations (up and down) *perpendicular* to their direction of movement. Because of being transversal when they are reflected through Iceland crystal the latter's internal structure separates the vibration into perpendicular directions, thus the emerging light is polarized at right angles to each other.

Fresnel was even able to rebut the main optical evidence that had convinced Newton of the superiority of the corpuscular theory: the sharp outline of shadows cast by large objects when deflected by light. Fresnel argued that one can explain the sharp outline as due to the large object's obstruction of certain waves at the edge of their propagation, but if one reduces the size of the object to the magnitude of the light wave then the light bends around the object as sound waves do. Finally another crucial test could be made based on the change of the velocity of light when passing through a lesser to a denser medium.

Newton had predicted that on the corpuscular theory the greater gravitational attraction of the denser medium would cause an acceleration of the light particles while on the wave theory the diffraction of the light would cause a retardation of the velocity. In a series of ingenious experiments now cited as *experimentum crucis* (critical experiments) begun in 1850, by French physicist Jean Léon Foucault confirmed that water or glass impedes the velocity of light in accordance with the wave theory. Then, based on these results, another French physicist named Hippolyte Louis Fizeau determined the velocity of light to be 300,000 kilometers per second, or 186,281 miles per second.

Illustrating how the correct paradigm of scientific inquiry leads to the determination of the relative truth of hypotheses, along with opening up new vistas of discovery, by the middle of the nineteenth century, Huygens's wave theory of light had superseded Newton's corpuscular theory, although radically new interpretations were yet to come, including the discovery that light was a form of electromagnetism, and the twentieth-century discovery that it can exhibit either wave or particle properties depending on the experimental conditions.

As for the discovery of electromagnetism, since ancient times electricity and magnetism were considered separate phenomena. But then in the winter of 1819–1820 Hans Christian Oersted (1777–1851), professor of natural philosophy in Copenhagen, during a course of lectures wondered if an electric current might have an effect on a magnetic needle. To test the supposition he placed an electrified wire at a right angle to the north south axis of a compass to no effect. Deciding to align the electrified wire parallel to the N-S axis he was surprised to find it produced a pronounced deflection of the needle, showing a relation between electricity and magnetism.

This was supported by Michael Faraday (1791–1867), a bookbinder's journeyman who apprenticed at age thirteen and therefore had little formal education but became an outstanding scientist, again showing the more common backgrounds of these later scientists. Attracted to science, he began attending the lectures by Sir Humphrey Davy at the Royal Institution in London becoming a member in 1823 and then a fellow of the Royal Society the following year. In 1833 he attained the position of Fullerian Professor of Chemistry at the Royal Institution. By then his reputation was such that he was offered knighthood and the presidency of the Royal Society but declined both. He is especially noted for his discovery of electromagnetic induction.

It had long been known that when iron filings were spread on a sheet of paper and a magnet placed underneath, the filings became aligned in a curved pattern around the magnet that, according to Sir Edmund Whittaker, "suggested to Faraday the idea of *lines of magnetic force*, or curves whose direction at every point coincides with the direction of the magnetic intensity at that point. . . ."[47] He then discovered that a moving magnet brought near an electric circuit induced a current, just as Oersted had found that an electric current changed the magnetic direction of the compass needle. As Whittaker continues:

> Faraday found that a current is induced in a circuit either when the strength of an adjacent current is altered, or when a magnet is brought near to the circuit, or when the circuit itself is moved about in presence of another current or a magnet. He saw from the first that in all cases the induction depends on the relative motion of the circuit and the lines of magnetic force in its vicinity. The precise nature of this dependence was the subject of long-continued further experiments. (p. 172)

Faraday's realization that a magnetic field can induce an electric current combined with Hans Christian Oersted's complementary discovery led to the conception of an independently existing electromagnetism either as a field or a current due to the interactions, replacing the previous conception that they were fluids. Reinforcing again the importance of mathematics in modern science, James Clerk Maxwell (1831–1879), based on these experimental discoveries, formulated intriguing equations describing the structure of electromagnetic fields and how they change in time due to the interactions. It was Maxwell's equations that implied that the velocity of the propagation of the waves of the electric field is identical to that of light indicating that light too is a form of electromagnetism.

This was confirmed towards the end of the nineteenth century when Heinrich Hertz (1857–1894) experimentally proved the existence of electromagnetic waves having the same velocity as that of light. Because electromagnetism involves the interaction of contiguous fields rather than forces emanated by discrete physical bodies in space as in Newtonian science, the conceptual framework of electromagnetism represents the beginning of the third scientific revolution that transformed our conception of reality. In their book *The Evolution of Physics*, Albert Einstein and Leopold Infeld declared that the "theoretical discovery of an electromagnetic wave spreading with the speed of light is one of the greatest achievements in the history of science."[48] Indeed, it was these developments that made possible the later introduction of radar, electric power, telegraphy, radio, television, the internet, and so forth. As Carl Sagan states in *The Demon-Haunted World: Science as a Candle in the Dark*, this "has done more to shape our civilization than any ten recent presidents and prime ministers" (p. 390).

Chapter V
THE ORIGINS OF CHEMISTRY AND MODERN ATOMISM

Since modern atomic theory along with celestial mechanics represent the two most significant theoretical developments in the physical sciences that changed our conception of the modern world from the ancient Aristotelian model to the modern mechanistic one, the latter requires a separate chapter. Recall that it was the ancient Greeks who first endeavored to understand the universe in a more empirical-rationalistic manner to replace the earlier mythical or theogonic interpretations. This required describing the primal elements from which everything arose, along with explaining how the diversity of nature came to be from this primal state.

Though it was Empedocles' conception of the four elements of fire, air, earth, and water as primary that was adopted by Aristotle, which prevailed throughout most of the past, modern classical science reinstated the atomic theory of Leucippus and Democritus, along with the theory of infinite particles composing the universe introduced by Anaxagoras and adopted by Epicurus and Lucretius. Thus it was natural philosophers like Mersenne, Galileo, Gassendi, Descartes, Boyle, Locke, and Newton who revived the atomic or particle theory in the seventeenth century by adopting the corpuscular-mechanistic framework, though the conception at that time was still entirely speculative and elementary.

Although as early as the third century BCE Anaxagoras had

declared that basic particles were infinitely divisible, when Newton adopted the corpuscular theory as the basic physical reality these particles were still mainly defined in terms of the Democratean primary qualities of solidity, shape, indivisibility, and motion (although Epicurus claimed they were composed of an inseparable minima), along with the more recent additions of mass, momentum, inertia, and gravitational attraction. Although the pseudosciences of alchemy and astrology were still pursued, the former by such distinguished natural philosophers as Boyle and Newton, they would soon be eclipsed by advances in modern classical science whose superior methodology led to the discovery of more elementary particles such as the electron and proton and an explanation of chemical compounds and reactions according to their exact molecular components, structures, and properties, rather than by God's will.

As was true of the transformation of the former notion of the celestial world to the modern conception of a gravity driven planetary and stellar universe according to mathematically defined astronomical laws, this new atomic and particle physics also would require a radical conceptual revision. Though not the first to use the balance to weigh the exact quantities of the reagents and products of chemical reactions, Antoine Laurent Lavoisier is considered the father of modern chemistry owing to his precise weighing of the components of combustion and oxidation that enabled him to determine that oxygen was a gas facilitating combustion, thereby refuting the prevailing phlogiston theory that postulated a fire-like element within combustible bodies. English chemist John Dalton similarly is regarded as the founder of modern atomism based on his discovery that natural elements like water, gases like carbon dioxide, and chemical compounds like sulfuric acid have a molecular structure that can be analyzed into specific atoms that combine according to simple numerical ratios according to their numbers: H_2O, CO_2 and H_2SO_4 respectively.

As religiously and rationally significant as was the transformation of the conception of a heavenly or celestial cosmology to a natural physical universe, for most of us, except for weather predictions and hurricanes and tornados, it is somewhat remote from our daily lives. This, however, is not true of the empirical sciences such as physics, chemistry, biology, physiology, medicine, engineering, etc. It is these sciences in particular that have radically changed our lives from what they were before the advent of science.

It was the overthrow of the theory that combustion was due to the expelling of phlogiston and replaced with the burning of oxygen that is usually credited with having been the major factor in the development of chemistry. It began with German physician Johann Joachim Becher's claim that combustion involved the burning off of the "fatty earth" described in his treatise *Physicae subterraneae* in 1669. Then German chemist George Ernest Stahl, in his book *Fundamenta Chymiae* (Fundamental Chemistry) in 1723, renamed Becher's *terra pinguis* "phlogiston," claiming that it was "the matter and principle of fire," though not fire as such. According to the phlogiston theory certain substances, like wood, charcoal, and phosphorus contain large amounts of this "*in*flammable principle" that they give off when heated that is combustion.

Then the Swedish chemist Carl Wilhelm Scheele, in a book translated as *Chemical Treatise on Air and Fire* published in 1777, reported his discovery that air consisted of two components: one highly flammable that he called "Fire air" and the other inflammable designated "Foul air," the first later renamed "oxygen" and the second "hydrogen." Though he detected a flammable substance in the air he did not investigate it. Like Scheele, Joseph Priestley in different experiments noticed that when substances are burned in air and the residue and the air are carefully weighed the residue usually *gained* weight while the volume of air *decreased,* contrary to the phlogiston theory that claimed the burning material *gives off* phlogiston and thus should weigh *less*, while the air *gaining* the phlogiston should weigh *more*.

Priestley's description of the transformation of calces (the residue of a burnt mineral) into metals is just one of many examples.

> For seeing the metal to be actually revived, and that in considerable quantity, at the same time that the air was diminished, I could not doubt, but that the calx was actually imbibing something from the air; and from its affects in making the calx into metal, it could be no other than that to which chemists had unanimously given the name of *phlogiston*.[49]

Thus Priestley is credited with discovering that it was "something from the air," a "new air," that caused the combustion, but as his final word "*phlogiston*" indicates, he was so committed to the phlogiston theory that he "concluded that the new gas must contain little or no phlogiston, and hence he called it *dephlogisticated air*" (pp. 126–27), which meant air that is free from phlogiston or the element of inflammability.

And so the honor of explaining the significance of the discovery is attributed to Lavoisier. After many failures to explain the process of combustion, it was at a dinner meeting in Paris with Priestley in 1774 that the solution occurred to him. Priestley "told Lavoisier at dinner of his discovery of dephlogisticated air, saying he 'had gotten it from *precip* [of *mercurius calcinatus*] *per se* and also *red lead*'; whereupon, he says, 'all the company . . . expressed great surprise'" (pp. 126–27). What caused the surprise was that the so-called dephlogisticated air produced by heating mercury oxide had properties the opposite of carbon dioxide produced by heating charcoal: it supported burning and respiration and did not combine with lime and alkalis. Repeating the experiment Lavoisier obtained a gas purer than ordinary air which convinced him that while measuring the components of chemical reactions is crucial to chemistry, so is choosing the right experiments.

He read two papers describing his experiments on the oxide of mercury titled "On the Nature of the Principle which Combines with Metals during Calcination and Increases their Weight" before the French Academy of Sciences, the first on Easter 1775 and the second on August 8, 1778. Having initially decided that the gas produced in Priestley's experiment though purer than common air was still a form of common air, when he learned of Priestley's later experiment showing that when reacting with nitrous oxide it was more soluble in water than common air, he concluded that while it was a constituent of common air it was not identical to it! He thus considered it a gas that was *absorbed* in the conversion of metals to calces or oxides when burnt in air and *emitted* when the oxides themselves were heated. He named the new gas "oxygene." According to chemist J. R. Partington:

> In 1782 Lavoisier says Condorcet had proposed the name "vital air" for pure air, but in a memoir received in 1777 . . . and published in 1781, entitled "General considerations on the nature of acids and on the principles composing them", Lavoisier called the base of pure air the "acidifying principle" or "oxigine principle" (*principe oxigine*), which he latter changed to "oxygene"[. . .] . (pp. 131–32)

The publication of Lavoisier's *Traité de Chimie* (Treatise on Chemistry) in 1789 established the superiority of the explanation involving oxygen over that of phlogiston. This not only overthrew the phlogiston theory, it brought about a revolution in chemistry. It no longer was assumed that common substances such as air and water were irreducible, but indicated they were compounded of more basic elements that opened up a whole new world of research. Although Priestley never gave up the phlogiston theory himself, in his last book he graciously acknowledged Lavoisier's contribution.

> There have been few, if any, revolutions in science so great, so sudden, and so general, as the prevalence of what is now usually termed the *new system of chemistry*, or that of the *Anti*phlogistons. . . . Though there had been some who occasionally expressed doubts of the existence of such a principle as that of *phlogiston*, nothing had been advanced that could have laid the foundation of *another system* before the labors of Mr. Lavoisier and his friends. . . .[50] (italics in original)

Yet as consequential as the chemical revolution has been, there was an impending revolution even more effective in transforming the conception of physical reality, namely, the reconstruction of the atomic theory. The success in chemistry of the experimental identification of oxygen and explanation of its function in combustion convinced scientists of the possibility of discovering the inner corpuscular elements of all ordinary substances, such as air, water, acids, metals, etc., along with the properties that could explain their combinations and reactions.

Thus while the intellectual and technological levels at the time of Leucippus and Democritus were insufficient to promote advances in the atomic theory, that was no longer true. Newton's belief stated earlier "that God in the Beginning form'd Matter in solid, massy, hard, impenetrable, movable Particles, of such Sizes and Figures . . . as most conduced to the End for which he form'd them," though a misrepresentation of their origin and properties, had finally been vindicated.

In 1787 Lavoisier, Claude Louis Berthollet, Guyton de Morveau, and Antoine François de Fourcroy published a book entitled *Méthode de Nomenclature Chimique* (The Method of Chemical Nomenclature) that presented the first modern list of elements based on recent experimental discoveries. Then in 1799 Joseph Louis Proust introduced "the law of constant proportions" stating that any sample of a compound or molecular substance, such as salt, always contains its constituents, sodium and chlorine, in fixed ratios by weight reinforcing the belief in the constancy of the reagents and the regularity of the reactions. Yet despite the considerable experimental evidence that substances were composed of more basic elements in ratios determined by their weights, there still was no *scientific* explanation as to why or how. And since Newton's belief that they were caused by God was no longer adequate the search began for an explanation, another indication of the transformation in the conception of the external world and how to investigate it and how to understand it.

The person credited with initiating the explanation, John Dalton, like many of his predecessors, was a most unlikely candidate. Born to a Quaker family in the tiny, rustic Village of Eaglesfield, England (1766–1844), his father was a cottage weaver while his mother supplemented their income by selling writing materials. Unable to enter a private school, he attended the village schools in the neighborhood but had acquired a sufficient background that he was able to teach in the village school from the early age of twelve to fourteen. While teaching there he was fortunate to meet a wealthy Quaker named Elihu Robinson who, along with being educated in natural philosophy, especially meteorology, corresponded with Benjamin Franklin.

Having noticed Dalton's mathematical aptitude when he won a dispute in mathematics, Robinson began tutoring him in mathematics with Dalton always appreciating the kindnesses, good advice, and intellectual awakening that Robinson and his cultured wife had contributed to his early development. Then, when he was fifteen, he moved to Kendal to become assistant in a boarding school rising to the position of principal. During his spare time he studied Latin, Greek, French, mathematics, and natural philosophy.

During the twelve years he lived in Kendal he made the acquaintance of a more unusual benefactor, another Quaker by the name of John Gough who, despite being blind and suffering from epilepsy, owing to his wealthy, well-educated, and intellectual family was able to acquire a sound knowledge of the classics, physics, mathematics, botany, and zoology. Though nine years older than Dalton and considerably more advanced in his studies, when he learned of their common interests and Dalton's intellectual aptitudes, Gough became his close friend and academic mentor. His family having an excellent library and an extensive collection of scientific instruments, he shared these with Dalton who, in gratitude, served as his reader and amanuensis. As a result, Dalton became well schooled in "mechanics, [Newton's] fluxions, algebra, geometry, chemistry (including some French chemical writings), astronomy and meteorology . . ."[51] (brackets added).

Because of this close intellectual relationship, when Dr. Barnes of New College in Manchester wrote to Gough in 1793 (who had become a widely respected mathematician) seeking his suggestions in filling a position of professor of mathematics and natural philosophy at New College, Gough unselfishly recommended Dalton for the position, even though it would mean severing their very close association. When the position of tutor at New College was offered to him, Dalton readily accepted, partially because of his dissatisfaction with his teaching at Kendal and also because he foresaw a more promising future in Manchester, which was confirmed when he later described his life there as "very happy and fulfilling."

Moving to Manchester he was immediately welcomed by the eminent "Mancunians," as the patricians of Manchester were called, and elected to the prestigious Manchester Literary and Philosophical Society the following year. As author Elizabeth C. Patterson states:

> The association which Dalton began with the Manchester Literary and Philosophical Society in 1794 was to continue until his death in 1844. During this half century the Society would play a central role in his life and he in its. Before it he read one hundred and seventeen papers, of which fifty-two were printed.

> For forty-four years he served as an officer—first as Secretary, then as Vice-President, and as President. To think of either—the Society or the man—is to think of the other. (pp. 59–60)

It was hearing the lectures and witnessing the experiments of English physician Dr. Thomas Garnett at New College that aroused his interest in molecular chemistry.

His early investigations and publications were centered on meteorology, including the nature of water vapor and the composition of the air and whether its components consisted of a mixture, a chemical compound, or some other structure. Then in a series of four essays he presented his research conclusions regarding gases, meteorology, and chemistry, in the last essay declaring that he had independently discovered Jacques Charles's gas law that all gases at constant temperature will, with the same increase in temperature, expand equally. In tribute, Patterson declares that the "wealth of material in these four essays is extraordinary. Even today they are hailed as 'epoch-making' and as 'laying the foundations for modern physical meteorology'" (p. 94).

It apparently was these initial experiments of the solubility of gases in water (similar to those of Robert Boyle) that led to his crucial insight that each element was composed of characteristic atoms that would be possible to distinguish by their atomic weights. The first explicit statement of this is in a paper he read to the Literary and Philosophical Society on October 21, 1803, entitled "On the Absorption of Gases by Water and Other Liquids."

> The greatest difficulty attending the mechanical hypothesis arises from different gases observing different laws. Why does water not admit its bulk of every kind of gas alike [i.e., why are they not equally soluble in water]? This question I have duly considered, and though I am not yet able to satisfy myself completely, I am nearly persuaded that the circumstance depends upon the weight and number of the ultimate particles of the

> several gases. . . . An enquiry into the relative weights of the ultimate particles of bodies is a subject, as far as I know, entirely new; I have lately been prosecuting this enquiry with remarkable success.[52] (brackets in the original)

While other chemists were only investigating the *relative* weights in which the components of substances *combine*, such as the densities or weights of hydrogen and oxygen composing water, Dalton was the first to attempt to determine the relative weights of the *components themselves.* Appended to the paper was a "Table of the relative weights of the ultimate particles of the gaseous and other bodies," while the paper itself presented the basic features of his atomic theory at the time.

After giving a series of lectures in Edinburgh and Glasgow that were highly praised—which must have been very gratifying considering his humble origins—he began writing his great work, a *New System of Chemical Philosophy,* published in 1808. The chapter "On Chemical Synthesis" was particularly significant because it explicitly states his original thesis that every sample of a basic substance such as water contains "*ultimate particles*" that always "*are perfectly alike in weight, figure, etc.*" When one considers the various possibilities as to how the ratios of the constituent particles could be construed, one begins to appreciate the complexity of the problem he faced. The key, he believed, lay in determining the individual atomic weights of the elements composing the substances. However, one can only claim that water is H_2O rather than HO_2 or H_3O_4 if one knows not only their atomic weights, but also in what numerical proportion they combine. As he summarized the challenge:

> In all chemical instigations, it has justly been considered an important object to ascertain the relative *weights* of the simples which constitute a compound. But unfortunately the enquiry has terminated here; whereas from the relative weights in the mass, the relative weights of the ultimate particles or atoms of

> the bodies might have been inferred, from which their number and weight in various other compounds would appear, in order to assist and guide future investigations, and to correct their results. Now it is one great object of this work, to show the importance and advantage of ascertaining *the relative weights of the ultimate particles, both of simple and compound bodies, the number of simple elementary particles which constitute one compound particle, and the number of less compound particles which enter into the formation of one more compound particle.* (p. 229)

Frank Greenaway states Dalton's "rules of greatest simplicity" as his guide to achieving this.

> 1 atom of A + 1 atom of B = 1 [compound] atom of C, binary.
> 1 atom of A + 2 atoms of B = 1 [compound] atom of D, ternary.
> 2 atoms of A + 1 atom of B = [compound] atom of E, ternary.
> 1 atom of A + 3 atoms of B = 1 [compound] atom F, quaternary.
> 3 atoms of A + 1 atom of B = 1 atom of G quaternary. etc. etc.[53]

Greenaway then presents the general rules that would guide Dalton in his "investigations respecting chemical synthesis": that is, how the elements of compounds were arranged.

> The composition of any substance must be constant (Law of Constant Composition). If two elements A and B combine to form more than two compounds then the various weights of A which combine with a fixed weight of B bear a simple ratio to one another (Law of Multiple Proportions). If two elements A and B combine separately with a third element C, then the weights of A and B which combine with a fixed weight of C bear a simple ratio to each other (Law of Reciprocal Proportion or Law of Equivalents). (p. 133)

He then assigned symbols to each of twenty elements with their atomic weights relative to hydrogen taken as 1 that he presented as a table of ELEMENTS in his *New System of Chemical Philosophy*. Although his symbols were later replaced by the Swedish chemist Jöns Jakob Berzelius (1779–1848) to those with which we are now familiar, he took the initial letter or letters from the Greek or Latin names or those rendered in English to designate the element: for example, "Fe (from the Latin *ferrum)*" to designate iron and H for hydrogen, O for oxygen, and Cl for chlorine, along with using *superscript* numerals to indicate the number of elements in a molecular compound such as water (H^2O).[54] It was Dalton who attempted the first systematic classification of atomic elements according to their atomic weights.

As Berzelius wrote when he first learned of Dalton's atomic hypothesis: "supposing Dalton's hypothesis be found correct, we should have to look upon it as the greatest advance that chemistry has ever yet made in its development into a science."[55] Later, after having read his *New System of Chemical Philosophy*, he wrote to Dalton that the "theory of multiple proportions is a mystery but for the Atomic Hypothesis, and as far as I have been able to judge, all the results so far obtained have contributed to justify this hypothesis" (p. 249). This was a vindication, finally, of the ancient theory of atoms introduced by Leucippus and Democritus that has proven so successful.

Yet despite Dalton's achievements, his rules of atomic composition were still suppositional based on atomic weights that did not provide the exact ratios in which the elements combine to form compound substances. In 1809, the year after Dalton published his great work, French chemist Joseph Gay-Lussac published a classic paper, "Memoir on the Combination of Gaseous Substances with Each Other," presenting his discovery known as "Gay-Lussac's Law of Combining Volumes" that measured the ratios of combining gases by their volumes, rather than their weights as Dalton had done, and also provided more exact proportions thus allowing a more precise determination of the ratios of the combinations of the atoms in their molecular structures, such as CO_2 or NH_3. As quoted by Leonard K. Nash:

> Thus it appears evident to me that gases always combine in the simplest proportions when they act on one another; and we have seen in . . . all the preceding examples that the ratio of combination is 1 to 1, 1 to 2, or 1 to 3. It is very important to observe that in considering weights there is no simple and finite [integral] relation between the elements of any one compound. . . . Gases, on the contrary, in whatever proportions they may combine, always give rise to compounds whose elements by volume are multiples of each other. (p. 260)

Gay-Lussac believed that his exact measurements of the integral ratios of combining gases were "very favorable" to Dalton's rules of combining weights, adding empirical support. Yet Dalton had strong objections, especially to Gay-Lussac's assumption that equal volumes of all gases under the same conditions contain equal numbers of atoms, instead maintaining that the different solubilities in water of the same volume of various gases implied that they were composed of atoms of different sizes and thus the same volume of different gases under the same conditions of temperature and pressure could not contain the same number of atoms.

If the density, mass, or weight of a substance is defined by the quantity of mass per unit volume, and the mass itself consists of the number and size of the particles composing the substance, then if two volumes of gas are equal but differ in density then this implies that either the number of particles are different or their sizes are or both. Thus Dalton concluded "that there are different numbers of particles in equal volumes of different gases was powerfully supported by experimental data on gaseous densities and combining volumes in gaseous reactions" (p. 266).

In addition he proposed as a maxim that the atoms in different gases vary in size: "*That every species of pure elastic fluid has its particles globular and all of a size; but that no two species agree in the size of their particles,* the pressure and temperature being the same" (p. 267). Thus Dalton attributed Gay-Lussac's claim that his Law of Combining Gases provided a more exact method for calculating the number of atoms in a substance to the *inexactness* of his experiments, despite the fact that Gay-Lussac had the reputation of being one of the most exacting scientists. Yet there was no agreement and so the search for additional evidence went on.

In an effort to resolve the problem and also to combine both Dalton's law of combining weights and Gay-Lussac's law of combining volumes, Italian chemist Amedeo Avogadro in 1811 published his "Essay on a Manner of Determining the Relative Masses of the Elementary Molecules of Bodies, and the Proportions in which They Enter into These Compounds" in the *Journal de Physique* (Journal of Physics). As he stated, if the simple or elementary gases rather than being monatomic or composed of single atoms were composite: if the "'particles' present in the gaseous elements *do not consist of the individual atoms* of the elements *but of groups of atoms of the same element joined in a single molecule* of that element" (p. 284), then the anomalies can be eliminated. Thus the original polyatomic atoms of the combining gases could separate and recombine into such proportions as to maintain a constant number in every volume of gas.

There were, however, three weaknesses in the theory. First, there was as yet no empirical evidence to support the theory. Second, even if the smallest particles of the volumes of combining gases were polyatomic there was no way of determining in what proportion they combined nor the nature of the combining force. Third, there was a conflict in assuming that the force binding the polyatomic substances was attractive and the fact that the gas pressure was repulsive. So the challenge remained to find a method to determine the exact number and ratios of the elements of compound or molecular particles.

In the succeeding years much of the research was directed at trying to solve this problem. When Count Alessandro Volta, an Italian physicist, in 1800 developed a voltaic pile or electrochemical battery that produced a continuous electric current several physicists who believed that the force binding the polyatomic structure might be electrical realized that Volta's electric current might decompose them into their components. Then in 1807 Sir Humphry Davy of the Royal Institution declared:

> If chemical union be of the [electrical] nature which I have ventured to suppose, however strong the natural electrical energies of the elements of bodies may be, yet there is every probability of a limit to their strength; whereas the powers of our artificial instruments seem capable of indefinite increase . . . [Consequently, we may] hope that the new [electrical] method of analysis may lead us to the discovery of the *true* elements of bodies. (p. 296)

The theoretical assumption was that if the monatomic particles in polyatomic elements such as ammonia or water have contrasting electrical charges causing them to bind, then connecting two terminals or electrodes to them that in turn were connected to the oppositely charged poles of Volta's battery with the charges strong enough, they would overcome the binding power of the charged particles thereby attracting them to the oppositely charged terminals. Today this would be called "decomposition by electrolysis." Within a year Volta had confirmed his theory by decomposing alkali metals by electrolysis.

But it was J. J. Berzelius, previously mentioned in connection with the symbolic naming of the elements, who discovered in the electrolysis of water that when two oppositely charged electrodes connected to the opposite terminals of a battery were inserted into the water it decomposed into the elements of oxygen and hydrogen owing to their being attracted to the oppositely charged electrodes. Finding this to be true of other compounds and inferring that it was due to the opposite charges of the monatomic elements, he proposed the dualistic classification of "electropositive" and "electronegative." Moreover, since electrical experiments had shown that like charges repel while opposite charges attract he inferred, as did Davy, that the stability of the elements could be explained by the attraction of their opposite charges which, when neutralized in polyatomic substances, left them uncharged.

Following his successful electrolysis of water, with the aid of Gay-Lussac's law of combing volumes, Berzelius was able to designate the correct molecular structure of compound substances using superscript numerals (later changed to subscripts) such as water (H^2O), ammonia (NH^3) nitrous oxide (N^2O) formed from gaseous elements. Though restricted to gases, his procedure was a significant advance because he was able to derive more precise atomic weights of the separate elements composing polyatomic substances. As a result, he spent more than a decade measuring the combing weights of the elements forming the compounds and publishing the results in tables in 1814, 1818, and 1826. And as indicated, having assigned symbols to each monatomic element based on the initial letter or letters of its assigned name, he then listed what he found to be the atomic structure of the polyatomic or compound substances with superscript numerals, indicating the ratios of their elements followed by what he had determined to be the compound substances' atomic weights.

Though this was a great improvement over the table of ELEMENTS published by Dalton, it still was not a sure method for determining the precise atomic weights of most elementary atoms nor the exact molecular structure of compound elements. The next major effort was the specific heat method of Alexis Petit and Pierre Dulong presented in a paper to the French Academy of Science on April 12, 1819. Aware that despite the advances in determining the exact proportions of the monatomic elements composing the polyatomic substances the results were still based on somewhat arbitrary principles, Petit and Dulong believed that discovering additional exact properties of the combining elements would enable a more precise computation. Having devised an exact experimental method for determining the *specific heats* of various elements (defined at the time as the degree of heat required to raise the temperature of the weight of a given substance by one degree relative to that of water), they believed they had found such a property. Known as the Petit and Dulong Law that the monatomic structures of all the compound substances have exactly the same capacity for heat, this enabled them to calculate the approximate values of the atomic weights. And since the molecular formulas are determined by the interrelation of atomic weights and combining weights this provided another way of determining the molecular structures. In their judgment, "Whatever may be the final opinion adopted with regard to this relation, it can henceforth serve as a control of the results of chemical analysis. In certain cases it may even offer the most exact method of arriving at information about the proportions of certain combinations" (p. 307).

Using the Petit Dulong Law and his own method based on gas densities and their combining volumes, along with the analogous behavior of the elements in chemical reactions and in the structures of crystals, according to Nash, Berzelius by

> judicious selections from among the various possibilities . . . he had, by 1840, arrived at atomic weights and molecular formulas that are in most cases in excellent agreement with those we now accept as correct. But alas, by this time a flood tide of skepticism was already lapping around the foundation of the atomic theory, and Berzelius' fine work did not receive the attention it deserved. (p. 309)

There was a discrepancy between Avogadro's belief that the relative weights of the heterogeneous particles could be inferred from the gas densities and the Petit Dulong Law that could not be applied to the gaseous elements.

Then in 1827 the French chemist J. B. A. Dumas devised a method for determining the gas densities at much higher temperatures that enabled him to study a much greater variety of substances at that higher temperature which in turn allowed him to reconcile the combining weights data derived from the Petit-Dulong Law with the relative gas densities obtained from Avogadro's hypothesis that equal volumes of gases do not contain equal numbers of particles. Briefly, the reconciliation could be achieved if, in addition to agreeing there were

> polyatomic molecules of the elements . . . it would now have to be further conceded that the polyatomic molecules of the different elements contain different numbers of the respective atoms . . . add[ing] the inability to explain why the molecules of different elements contain different numbers of their respective atoms. (p. 312; brackets added)

It took a little more than a quarter of a century before this phase of atomic physics reached a resolution by the Italian chemist Stanislao Cannizzaro. During that time there continued to be new discoveries despite the prevailing skepticism to accepting the truth of the atomic theory, such as the kinetic explanation of gas pressure as due to the mobility of dispersed particles in the gas and new evidence to support Avogadro's theory of polyatomic particles. Accepting Avogadro's law that equal volumes of similar gases contain the same number of particles of which some were polyatomic, Cannizzaro concluded that it should not be assumed that equal volumes of *different* gases contain the same number of basic particles. But if that were true, the weight of the atoms could not be inferred unless it was known how many atoms were contained in the volume, a near impossibility.

Thus he introduced a different procedure that involved

weighing the densities of *various* compound gases containing the same element. Knowing the densities per unit volume of a number of compound gases containing that element, the weight of the element could be determined by what fraction of the weight of the compound was due to that element. Beginning with the smallest ratio, he found that in succeeding weightier compounds the ratios of that element were always in whole numbers. He then realized that he could calculate the *relative weights* of the elementary particles by comparing their ratios in the weights of the various compounds.

If the elementary compound contains one atom of that element this would give the atomic weight of that element. Then following Berzelius' convention that established the atomic weight of hydrogen as the standard of 1, the weights of the other elements relative to hydrogen could be assigned: carbon as 12, oxygen as 16, sulfur as 32, and so forth. As these atomic weights agreed with those derived by the method of specific heats used by Petit and Dulong, this provided strong confirmation of the theory of atomic weights. Thus a half century later thanks to the efforts of preceding experimentalists, Cannizzaro proved Dalton's belief in 1808 of "'the importance and advantage of ascertaining *the relative weights of the ultimate particles of both simple and compound bodies*'" (pp. 318–19).

About two decades after Cannizzaro's generally accepted determination of the atomic weights, in 1848 his research culminated in the independent publication respectively of the Periodic Law by Julius Lothar Meyer and the Periodic Table by Dmitri Ivanovich Mendeleev. Mendeleev's table was published in Russian in April 1869 and though Meyer's paper containing his Periodic Law was dated December 1869, it was not published in Germany until 1870. In a Faraday Lecture given to the Fellows of the Chemical Society of the Royal Institution in 1889 Mendeleev gave a succinct but comprehensive summary of what had been achieved up to that time and what could be anticipated in the future.

1. The elements, if arranged according to their atomic weights, exhibit an evident *periodicity* of properties.
2. Elements which are similar as regards their chemical properties have atomic weights which are either of nearly the same value (e.g., platinum, iridium, osmium) or which increase regularly (e.g., potassium, rubidium, caesium).
3. The arrangement of the elements, or of groups of elements, in the order of their atomic weights, corresponds to their so-called *valences* [the combining power of an element] as well as, to some extent, to their distinctive chemical properties—as is apparent, among other series, in that of lithium, beryllium, barium, carbon, nitrogen, oxygen, and iron (brackets added).
4. The elements which are the most widely diffused have small atomic weights.
5. The *magnitude* of the atomic weight determines the character of the element, just as the magnitude of the molecule determines the character of a compound.
6. We must expect the discovery of many yet *unknown* elements—for example, elements analogous to aluminum and silicon, whose atomic weight would be between 65 and 75.
7. The atomic weight of an element may sometimes be amended by a knowledge of those of the contiguous elements. Thus, the atomic weight of tellurium must lie between 123 and 126, and cannot be 128.
8. Certain characteristic properties of the elements can be foretold from their atomic weights.[56]

This enabled Mendeleev in 1895 to present his Periodic Table consisting of two columns, one vertical and the other horizontal. Both columns listed the elements under the headings given by Berzelius, with the vertical column presenting them according to their atomic weights and chemical properties while the horizontal column listed them in increasing numerals according to their atomic numbers as determined by the number of protons in their nucleus, beginning with hydrogen as 1 since it contains 1 proton. As now written, it appears as H but in a water molecule as H_2O (with the *subscript* $_2$) because it contains two elements of hydrogen.

What a great progress had been made by chemists since Lavoisier's explicit recognition of oxygen as the combustable gas and the resurrection of the ancient atomic theory of Democritus and Leucippus by Dalton culminating in Mendeleev's Periodic Table. We now await the discovery of new elements and their molecular structure in compounds along with the interior structure of the atom.

Chapter VI

TRANSITION TO THE THIRD REALITY IN THE LATE NINETEENTH AND TWENTIETH CENTURIES

We have just seen how long it had taken and how arduous the task was to replace the Aristotelian and religious worldviews (though religionists still do not accept the fact that there is no longer any plausibility to religious explanations) with the early modern scientific corpuscular-mechanistic explanation of the universe and physiological nature of human beings. Starting with Copernicus in the middle of the sixteenth century, it was not until the latter half of the nineteenth and mainly in the twentieth century that this earlier, more preliminary mechanistic conception of reality was replaced by the third major transition, along with the consequent technological advances.

As American physicist Michio Kaku states in his latest book, "the time was right for the emergence of an Einstein. In 1905, the old physical world of Newton was crumbling in light of experiments that clearly suggested a new physics was about to be born, waiting for a genius to show the way."[57] Though a continuation of the earlier scientific discoveries and theoretical advances of the previous three centuries, it brought about a decisively greater understanding of the world and of human beings due to a much more realistic and extensive conception of the physical universe owing to greatly improved telescopic observations and spectroscopic evidence; deeper probing of the interior structure of the atom; the rejection of Newton's deterministic-mechanistic universe of absolute space and time due to the uncertainty principle in quantum mechanics and Einstein's view of space-time as a four-dimensional field in his general theory of relativity; paradoxes such as the wave-particle duality; and the replacement of "intelligent design" and "special creation" of humans with the confirmation of evolution, brain research (eliminating the soul), and the deciphering of the genetic code.

Ernest Rutherford, the father of nuclear physics, "once said that the rapidity of advance during the years 1895–1915 has seldom, if ever, been equaled in the history of science." Even if true at that time, Rutherford did not live to witness Hubble's astonishing telescopic discoveries and evidence of the inflationary recession of the universe, Georges Lemaître's anticipation of the Big Bang explanation of the origin of the universe, tunneling microscopes, nuclear fission and fusion, nuclear reactors and particle accelerators, computer technology, genetic decoding and engineering, along with the lunar landing and controlled interplanetary space flights. One can truthfully say there were more new advances in science in the twentieth century than in the entire past history of science that emended Newton's deterministic, absolutistic, mechanistic worldview with a much more extensive and complex conception of reality.

As Einstein and Infeld wrote in 1951:

> During the second half of the nineteenth century new and revolutionary ideas were introduced into physics; they opened the way to a new philosophical view, differing from the mechanical one. The results of the work of Faraday, Maxwell, and Hertz led to the development of modern physics, to the creation of new concepts, forming a new picture of reality.[58]

As we have seen, some of the major changes in physics involved the detection of diffraction patterns supporting the wave theory of light to complement Newton's corpuscular theory; the replacement of the independence of electricity and magnetism with the concept of an electromagnetic field and the recognition that light, too, is an electromagnetic phenomenon; Einstein's introduction of a variable four-dimensional space-time field to replace Newton's conception of gravity as a mutually attractive force between objects; the rejection of indivisible solid atoms for a largely space filled atom conceived in terms of an inner solar structure of subatomic particles with electrical charges that determine the atoms' properties and interactions.

This also included the later discovery that whether light reacts as waves or corpuscles depends on the experimental conditions, thus introducing the wave-particle duality of light; the rejection of the ether as a necessary medium for the transmission of light and gravity; the fact that spatial or temporal measurements are not absolute but relative to the velocity of the measurer and its effect on the measuring device; the discovery of quanta of energy and the uncertainty of the measurements and existence of subatomic particles due to the indeterminate nature of their properties until measured; the red-shift in the light waves from outer space indicative of the recession of the light source and of an expanding universe, along with evidence that at least *our universe* began with a big bang about 13.7 billion years ago; string theory to replace all other theories although as yet there is no evidence to support it; and the possibility of a pluralistic universe with different laws of which ours is only a miniscule extension. It seems there is no end to the future discoveries facing science.

While it is said that "the eighteenth century is the century of Newton," it is equally true that it was in that century that scientists began the critical task of questioning and testing the fundamental concepts of the Newtonian scientific framework, a crucial function of science that accounts for its success and advancement. For unlike the followers of Aristotle (but not Aristotle himself) who claimed that all knowledge could be found in his works, and religionists who still claim that divinely revealed scripture as interpreted by church authorities contains the ultimate, eternal truth and thus scientific knowledge has only an instrumental value, the obvious advances of science and the tremendous changes in our conception and relation to the world that it has brought about belie this. The task now is to describe these later developments as clearly and simply as possible.

Among the first challenges to Newton's system was his conception of the corpuscular theory of light based on his prismatic discovery of the rays of light and the sharp boundaries of shadows. At the time there was considerable controversy over whether light consisted of waves or of corpuscles and whether it had a finite velocity or was propagated instantaneously. Most continental natural philosophers, such as Hooke and Huygens, defended the wave theory. It was Newton and Olaus Roemer who first succeeded calculating the time for light to reach the earth from the sun to be about eight minutes. Hooke had proposed that light is a form of wave motion in a medium that was supported by Huygens in a lengthy article published in 1690 titled *Traité de la lumière où sont expliquées les causes de ce qui arrive dans la réflexion et dans la réfraction. Et particulièrement dans l'étrange réfraction du cristal d'Islande* (Treatise on light explaining the cause of reflection and refraction. And particularly in the strange refraction of Island crystal). But while Huygens's extensive research was extremely influential on the continent, Newton's prestige in the British Commonwealth was such that his corpuscular interpretation of light was dominate there until it was finally contested in the beginning of the nineteenth century.

As we now know, Thomas Young was among the first to challenge the corpuscular theory in a paper entitled "Outlines of Experiments and Inquiries Respecting Sound and Light" published in the *Philosophical Transactions of the Royal Society* in 1800.

Impressed by the similarity of light and sound, he favored the theory that light is transmitted by waves. Along with additional criticisms of the corpuscular theory, his major objection was based on experiments showing that light radiated through small apertures produced diffraction patterns, which offered clear evidence of waves. In a paper titled "On the Theory of Light and Colours" in 1802, he described the light and dark bands in the diffraction patterns in terms now known respectively as constructive (in phase) and destructive (out of phase).[59]

Despite the nearly incontrovertible evidence, the respect for Newton was so strong that the reaction in England was hostile rather than approving. However, that changed when a gifted French engineer named Augustine Jean Fresnel, having had doubts about the corpuscular theory and impressed by the results of Young's experiments, decided to test whether Newton's generally accepted explanation was true. His explanation was that the wave bands were produced by the attractive or repelling effects of the shapes and thickness of the apertures on the light corpuscles as they passed through. But when Fresnel, based on experiments that varied the shapes and thickness of the tiny openings, found they had no effect on the diffraction patterns that depended solely on the size of the aperture, even the English were converted. When he presented the results of his experiments in a "Memoir on the Diffraction of Light" to the Paris Academy in 1819, he was awarded a prize.

It was Fresnel's ability to quantify his results and predict additional testable consequences that was particularly convincing. As an example, rejecting Newton's peculiar explanation of the polarization of light rays emerging from Iceland spar by attributing it to different sides of the corpuscles constituting the light rays, he concluded that the wave theory could provide a simpler and more plausible explanation if one accepted that the light waves were transverse, produced by the perpendicular vibrations upward and downward of the undulations emitted from the apertures. Thus Fresnel's explanation proved to be a final vindication of Hooke's and Huygens's advocacy of the wave conception of light.

Then Jean Léon Foucault, in addition to demonstrating the rotation of the earth by hanging a huge pendulum from the roof of the Panthéon in Paris in 1851 and inventing the gyroscope, in what is often referred to as an *experimentum crucis*, by exacting experiments begun in 1850 and confirmed in 1862, he demonstrated that light travels more slowly in denser media, thereby disproving Newton's prediction that it would travel faster due to the greater attractive force of the more dense media on the corpuscles. This was the final blow to the corpuscular theory because nothing is more fatal to a theory than a definite disconfirmation of a crucial prediction.

Furthermore, based on Foucault's experiments, Hippolyte Louis Fizeau determined the velocity of light to be 300,000 kilometers per second, thus settling the dispute as to whether light was transmitted instantaneously, as many had believed, or with a finite velocity. So by mid-century, owing to these ingenious experiments confirming the properties of light, most scientists accepted the wave theory over Newton's corpuscular theory. One would have to await Einstein's 1905 paper explaining the photoelectric effect in terms of discrete quanta of light before the particulate properties of light would again be accepted.

There is a further aspect to the discussion, the existence of the aether. Unlike the transmission of corpuscles, waves being a configuration of a medium required the existence of a medium that later was called "aether." Although in Newton's corpuscular theory light did not require a medium for its transmission, in the long quote in the previous chapter on Newton he did predict that in the future a unified theory would be developed to explain all the various astronomical, physical, chemical, optical, and even neurological phenomena in one integrated theoretical framework.

Yet nothing illustrates more clearly the difficulty that such transitions present than the fact that even someone of Newton's genius could not make a complete break with the older tradition. He continued to describe the underlying unifying force as due to the "vibrations" of an "a certain most subtle spirit"[60] that conflicted with his criticism of "feigned hypotheses" and astute definition of the correct scientific method. Nevertheless, the concept of an electric spirit required for the transmission of electricity and possessing the vibratory elasticity needed for the transmission of light waves would soon be replaced by an all pervasive subtle medium called the "aether," that in turn would be replaced later by Einstein's concept of a space-time field. Such are the contingent fortunes of scientific theories owing to the continuous experimental testing of past theories. But despite the almost universal acceptance of the aether theory, the discovery of electromagnetism, radiation, and the invariant velocity of light, the explanatory value of the aether became nil and was discarded after the Michelson-Morely experiments and Einstein's Special Theory of Relativity proved the invariant velocity of light.

Turning now to another scientific revolution that would contest Newton's mechanistic worldview, though it had been previously suspected that electricity and magnetism were not independent, it was Hans Christian Oersted in 1807 who demonstrated their interconnectedness. Investigating the effects of an electric current on the deflection of a magnetic needle, he discovered that *a changing current* induced a magnetic field and announced his discovery in an article titled *Experimenta circa effectum conflictus electrici in acum maneticam* in 1820.

Describing it as "the *conflict of electricity*," he wrote that the magnetic field produced by the electric wire occupied a curved space surrounding the magnet, whose direction was dependent upon whether the wire was above or below the needle along with the direction of the current. The French academician François Arago, upon learning of Oersted's experiments, devised his own experiment in 1820 showing that when a current is passed though a circular copper wire it attracts unmagnetized iron filings that remain attracted to the wire as long as it is electrified, but immediately fall off when the current ceases. Then in 1824 he reported that a rotated copper disk produced a similar rotation of a magnetic needle supported above it.

Another major contributor, Andre Marie Ampère, shortly after Oersted's publications, proposed the laws governing the electric current's deflection of the magnetic needle, along with the reciprocal attractions and repulsions of electric currents. His outstanding achievements are memorialized in the "well-known 'Ampère's Rule', formulated by him for determining the deflection of a magnet by an electric current, and in the *ampere*, the practical unit of electric current, which is named after him."[61]

Next in the succession of contributors is Michael Faraday, referred to by British historian of science Charles A. Singer as "one of the greatest of scientific geniuses," introduced the concept of a "field of force" to describe the magnetic aura created by a current from an electrified wire. Just as Ampère had demonstrated that a spherical coil of wire produces a magnetic attraction when a current is passed through it, Faraday in a series of experiments demonstrated the converse—that moving magnets could "induce" an electric current. Realizing that "the essential factor in the production of the magneto-electric effects was change, movement of the magnet or of the coil, or making and breaking of the current or the contact," he concluded that they are what produce the "fields of force" (p. 363). This discovery of electro-magnetic induction resulted in the invention of the dynamo and the production of electricity on such a large scale that it transformed civilization. We have become so dependent upon readily available electricity that whenever there is a loss of electricity in the Western world due to some natural disaster or snow storm life becomes chaotic.

While Faraday's great contribution was the experimental discovery of what generated the magneto-electric currents, it was James Clerk Maxwell who, by his equations describing the structure and changes of the "electromagnetic field," fully developed the theory of electromagnetism that had been confirmed by Heinrich Hertz at the end of the nineteenth century. In his work *On a Dynamical Theory of the Electro-magnetic Field*, published in 1864, Maxwell declared that since the electromagnetic waves were transverse analogous to light and had the same velocity as light, the latter must also be a form of electromagnetism. It was this that led Einstein and Infeld to declare that the "theoretical discovery of an electromagnetic wave spreading with the speed of light is one of the greatest achievements in the history of science."

The next advance was the experimental attempt to discover the interior structure of the atom. In our previous discussion of atomism it was pointed out how the recognition of oxygen's role in combustion by Lavoisier led to the development of chemistry and the identification of additional elements such as hydrogen and nitrogen, followed by Dalton's and Berzelius's determination, based on their atomic weights, of the ratios of the elements constituting the molecular structure of substances such as air and water. This culminated in Meyer's and Mendeleev's organization of the elements into the Periodic Table based on the periodicity of their properties according to their atomic weights. At the time, the assumption was that the monatomic elements were internally compact and therefore indivisible as was believed by Leucippus and Democritus, the originators of atomism. Their observable properties were explained by their sizes, shapes, mass, weight, and motion.

There had been some indications, however, that that this was not true. Even as early as the fourth century BCE, Epicurus had conjectured that the various sizes and shapes of atoms might be due to internal "minima," which, like present-day quarks, could not exist separately. And even Newton in Query 8 of the *Opticks*, had asked:

"Do not all fix's Bodies, when heated beyond a certain degree, emit Light and shine; and is not this Emission perform'd by the vibrating motion of its parts?" Newton also deduced that the produced light consisted of identifying emissions, a process now called spectral analysis or spectroscopy, that provides a more extensive and accurate identification of their unique chemical properties. Even though his mechanistic explanation—that it was the vibratory or oscillatory frequencies of the atoms or molecules caused by the heat that produced the emissions—was mistaken, his suggestion that heated objects or other substances produce identifying spectra introduced a new method for analyzing the properties of phenomena and opened up a whole new area of scientific enquiry.

One of the first to pursue the inquiry was Thomas Melville who discovered the spectrum of salt, while William Wollaston and Joseph von Fraunhofer initiated the science of astrophysics when, in 1802, Wollaston started examining the solar spectrum and Fraunhofer invented the spectroscope in 1814 for mapping the latter. Discovering that basic gases can also be identified by their signature spectrum, Anders Jöns Ångström observed the spectrum of hydrogen and, with Julius Plücker, computed the wavelengths of the four spectral emissions of hydrogen.

But it was Gustav Robert Kirchhoff and Robert Bunsen—the latter having invented the burners needed to produce a pure flame required for the spectral analysis, which is still used in chemistry laboratories today—who largely created the science of spectroscopy by definitely establishing in 1859 that each chemical substance emits its own signature spectrum. Even though the evidence was more complex than they originally believed, since at higher temperatures substances produce different line spectra, they realized that the method could be used to discover new elements to fill the gaps in the Periodic Table.

In a joint statement they described the purpose and nature of their "spectrum-analytic method."

> In spectrum analysis . . . the colored lines appear unaffected by . . . external influences and unchanged by the intervention of other materials. The positions occupied [by the lines] in the spectrum determine a chemical property of a similar unchangeable and fundamental nature as the atomic weight . . . with an almost astronomical accuracy. What gives the spectrum-analytic method a quite special significance is . . . that it extends in an almost unlimited way the limits imposed up till now on the chemical characterization of matter.[62]

Although spectral analysis would prove limited in physics and chemistry with the discovery of the inner structure of the atom, solar and stellar spectral analysis has provided our main information about the composition of the planets, comets, and other astronomical phenomena.

But along with spectral analysis, electrical developments before the discovery of subatomic particles provided additional understanding of electrical conduction and radiation and facilitated the discovery of the charged subatomic particles. For example, Michel Faraday constructed the cathode ray tube for investigating electrical discharges in gases. In vacuum tube experiments conducted between 1833 and 1838, he passed a current from a negative electrode or cathode to a positive electrode or anode through which rarified gases passed. This induced a glow on the inner surface of the opposite end of the tube.

The glow resembling that of phosphorescence, scientists were perplexed by what caused the luminosity. Because the effectiveness of the experiments depended upon the extent of the vacuum and the strength of the current, Johann Hittorf , Philipp Lenard, and Sir William Crookes performed numerous experiments to improve its effectiveness. Then Heinrich Rühmkorff introduced a better induction coil to generate higher voltage that produced stronger currents in the vacuum tube.

The reason for describing these experiments is their influence on later emission research. For example, intrigued by the previous experiments on cathode ray tubes, Wilhelm Röntgen, in repeating those experiments, discovered a ray with an amazing penetrating power. Experimenting in a dark room on November 8, 1895, having completely covered a Hittorf cathode tube with a black cardboard to block any rays, he suddenly noticed that a sheet of paper coated with barium platinum-cyanide, located a short distance from the cathode tube, fluoresced indicating that some rays must have penetrated the cardboard and activated the coating thereby producing the glow. He was especially astonished when he held his hand between the tube and the coated paper and saw that the rays penetrated his hand except where obstructed by the bones and a ring on his finger. He later produced a dramatic photograph of his skeletal hand during the announcement of his discovery of what he called X-rays, because of their mysterious penetrating power. For this work he was awarded the first Nobel Prize in physics in 1901.

As knowledge of Röntgen's remarkable discovery spread throughout the world, other physicists began investigating the phenomena. Learning that Röntgen believed that whatever caused the fluorescent glow on the inner surface at the end of the cathode ray tube also caused the X-rays, a year later Henri Becquerel began experimenting with phosphorescent substances to determine whether they emitted X-rays, but discovering they did not, he decided to use uranium salt. To test his hypothesis he proceeded like Röntgen by covering a photographic plate with black paper and placing the uranium salt on top, assuming that the heat from the light of the sun would activate the uranium and produce an image on the photographic plate. After leaving the arrangement in the sun for several hours and then developing the photographic plate, he did find an image of the uranium salt.

Intending to reproduce the experiment, Becquerel placed a copper cross between the wrapped photographic plate and the uranium salt, but finding the sky overcast he placed the copper cross in a closed cupboard for several days. Expecting the photo to be unexposed because of the lack of sunlight acting on the uranium salt, he was "stupefied" to discover an even sharper image of the cross than of the previous metals. Realizing that his assumption that the uranium salt had to be activated by light to produce the image was false, he concluded that the "rays" were generated spontaneously within the uranium compound itself. But though he was the first to identify radiation, his discovery was generally ignored until the research of Pierre and Marie Curie.

It was a year later in 1897 that Becquerel's discovery began to be accepted when Pierre Curie, director of the laboratory in the school of Industrial Physics and Chemistry in Paris, suggested to a Polish émigré Marya Sklodowska, who had enrolled as a graduate student, that for her doctoral dissertation she might investigate "Becquerel rays." It also resulted in Pierre and Marya getting married the following year and because of their mutual attraction, common interests, and collaboration they formed the most famous husband and wife team in the history of science, with Marie Curie (her married name) the most scientifically gifted of the two.

Later at her initiative, they began investigating Becquerel rays using Pierre's improved electrometer in place of Becquerel's electroscope to obtain a more precise measurement of the degree of radiation. Marie discovered that the source of the radiation was uranium atoms and introduced the term "radioactive substance" in the title of a paper in 1898 to describe the material emitting the spontaneous radiation. She also developed a technique for isolating radioactive isotopes and discovered two elements, polonium (named after her country of origin) and radium.

Marie and Pierre shared the Nobel Prize in physics with Becquerel in 1903 for their discoveries and she earned another Nobel Prize in chemistry in 1911, the first person to win two Nobel Prizes and one of only two people to have been awarded a Nobel Prize in two fields. Tragically, they both suffered dreadful deaths: Pierre at age forty-six when he was killed by a vehicle while crossing the Rue Dauphine on the left bank in Paris, and Marie at sixty-seven died in a sanatorium as a result of her exposure to radiation from which she suffered for many years. Conceivably their discoveries in radioactivity, which helped convince physicists of the limitations of Newtonian science due to the existence of a deeper level of physical reality, provides some consolation for their tragic deaths.

It is to the first discovery of an element at this deeper level that we now turn, and again it relates to experiments with the cathode ray tube. As was the case with light, there was a division between the English investigators and those on the continent as to the nature of cathode rays, the English adopted a particle interpretation and the Europeans a wave theory. As the Englishman William Crookes stated:

> As is well known there are two opposing views on the nature of the cathode rays. The earlier one . . . adopted by the English physicists, considers the rays as negatively-charged particles. According to the second one, more representative of the German physicists, especially Goldstein, Wiedemann, Hertz and Lenard, the cathode rays are processes . . . in the aether. (pp. 81–82)

In 1879 Crooks also claimed that "In studying this fourth state of matter we seem at last to have within our grasp . . . the little indivisible particles which with good warrant are supposed to constitute the physical basis of the Universe" (p. 80). But it was J. J. Thomson who determined that the cathode rays were particles, later called "electrons" after measuring their mass/charge ratio, a property of particles not waves. A year previous to Thomson's announcement of his experimental measurement, Emil Wiechert's University of Königsberg experiments on cathode rays indicated that they had a mass much lighter than hydrogen, a property of particles. He could have been the first to have discovered a subatomic particle, but for ideological reasons he desisted. As Abraham Pais states: "It is the first time ever that a subatomic particle is mentioned in print and sensible bounds for its mass are given" (p. 82).

It was in the following year, 1897, in April that Thomson presented to the Royal Society of London his experimental determination of the ratio of charge to mass (*e/m*) of cathode rays that definitely convincing him they were particles:

> On the hypothesis that the cathode rays are charged particles moving with high velocities [it follows] that the size of the carriers must be small compared with the dimensions of ordinary atoms or molecules. The assumption of a state of matter more finely subdivided than the atom is a somewhat startling one. . . . (p. 85)

In the following August he submitted an article describing his results to the *Philosophical Magazine.*[63] He asserted first that the particles were negatively charged, having measured it by an electrometer. Then in an experiment with a cathode ray tube in which he had inserted two separated parallel electrode plates of contrasting charge, attached oppositely to the center surface of the tube to generate an electric field perpendicular to the length of the tube, he directed the beam of cathode rays between the electrode plates, noting the rays' *electrical* deflection indicated by the displacement of the luminous glow in the enclosed glass globe at the opposite end. Repeating the experiment but generating a magnetic rather than an electrical field, he detected their *magnetic* deflection, another property of particles rather than waves.

From these experiments he was able to measure the charge-to-mass ratio, *e/m*, along with calculating the velocity of the particles and discovering that since they were identical in all the experiments they must be an elemental component of nature. As he concluded:

> On this view we have in the cathode rays matter in a new state, a state in which the subdivision of matter is carried very much

> further than in the ordinary gaseous state: a state in which all matter . . . is of one and the same kind; this matter being the substance from which all chemical elements are built up. (pp. 85–86)

This is a remarkable premonition in that as we now know the electron to be a basic particle of nature that accounts for the chemical properties and interactions of the elements.

Thomson finally confirmed his discovery of electrons in 1899 using the cloud chamber method developed by his student C. T. R. Wilson. In that experiment charged particles forming the nuclei of condensed droplets of supersaturated water vapor enabled him to measure separately the charge and mass of the particles. The charge he found to be $e \approx 6.8 \times 10^{-10}$ esu (electrostatic units) and the mass of the electron to be 3×10^{-26} g, a very accurate measurement for the time (p. 86). Thus electrification could be explained as the result of splitting the electron from the atom. While it was George Stoney in 1894 who introduced the term "electron" to symbolize the unit of electrical charge lost or gained when an atom or molecule acquired a net electrical charge, thus becoming an ion, it was Thomson's measurement of the electron's mass that is the reason he is considered "the discoverer of the electron." (p. 86)

Although this completes the account of the discovery of the electron it does not explain the curious reason why Wiechert, the person who, a year before Thomson, had experimentally detected that cathode rays appeared to be particles rather than waves because of their mass, was not credited with making the first discovery. The conceded explanation is that in contrast to England, where the existence of atoms and corpuscles had been commonly accepted among scientists since Newton, in Austria and Germany where the "positivist philosophy" of the Viennese physicist Ernst Mach prevailed, it was thought to be "unscientific" to accept the hypothetical existence of entities like atoms or electrons that could not be actually observed, but only inferred from the experimental evidence.

During the eighteenth century the Scottish philosopher David Hume in his critique of science denied the existence of insensible objects, while in Prussia Immanuel Kant had limited all cognition to the phenomenal world of sensory experience declaring that knowledge of imperceptible "things in themselves" was impossible. But while in England Hume's skepticism was offset by Locke's philosophical defense of the existence of the *insensible primary qualities* of microscopic objects and Newton's acceptance of the existence of unobservable atoms and corpuscles, the skepticism of Kant was perpetuated in Germany and Austria by the positivist philosophy of Ernst Mach (1838–1916). Influenced by his positivism, Wiechert did not believe or declare he had made a real discovery and therefore his priority is not acknowledged. Even when I was studying the philosophy of science Mach was still a strongly obstructive influence.

Having acquired a BS degree as an undergraduate and then being attracted to philosophy by the writings of the ancient Greeks and such contemporaries as Alfred North Whitehead, Hilary Putnam, Bertrand Russell, John Dewey, and Thomas Nagel, I had the misfortune of teaching during the second half of the twentieth century when positivism and ordinary language philosophy prevailed, both of which I considered superficial and ill informed, explaining why they no longer have *any* following. Who today would accept Kant's and Mach's positivistic thesis that scientific research is limited to the observable world or that ordinary language philosophy is preferable to the technical language of science as claimed by G. E Moore, Gilbert Ryle, and Ludwig Wittgenstein?[64]

Returning to the account of the investigations of atomic emissions that greatly influenced the transition from the Newtonian worldview to that of contemporary science, the next contributor was Ernest Rutherford who was born on the South Island of New Zealand where he received his early education. As an indication of his intelligence, on his entrance examination to Nelson College "he scored 580 out of a possible 600 points and was first in English, French, Latin, history, mathematics, physics, and chemistry."[65] After graduating from Nelson College, he received a fellowship to Canterbury College, where he earned an MA degree and then won a scholarship to Cambridge University in 1851, which enabled him to continue his education in England. As related by Emilio Segrè: "It is said that when the announcement of his prize arrived, Rutherford was on the family farm digging potatoes. He read the telegram bringing the news and said, 'This is the last potato I have dug in my life'" (p. 49). Thus began one of the most brilliant careers ever in experimental physics.

Arriving in London when he was twenty-four years of age in 1895, there could hardly have been a more auspicious time for him to begin his research. Röntgen would discover X-rays in the next month, Becquerel would detect uranium radiation in six months, and the Curies in about a year and a half would identify the existence of the radioactive substances polonium, thorium, and radium. Shortly after his arrival, Rutherford collaborated with J. J. Thomson, who immediately recognized his research talents. Rutherford measured the ionization produced by X-rays and, subsequent to its discovery by the Curies, uranium. From these experiments he learned that uranium radiation was not homogeneous but emitted two kinds of rays identified by their different emission capacities: *alpha* (α) being highly absorbable and *beta* (β) very penetrating. It was the *alpha* rays that were absorbed by the black cardboard wrapper in Becquerel's experiment and the *beta* rays that activated the coated screen.

Despite his close collaboration with Thomson at the Cavendish Laboratory, one of the greatest research centers in the world, when offered a professorship in 1898 at a much higher salary at McGill University in Montreal, Canada, Rutherford accepted. When he arrived at McGill he found newly built physics and chemistry laboratories and such distinguished colleagues as Frederick

Soddy and Otto Hahn. As a further reward, when John Cox, the chairman of the physics department, observed Rutherford doing his research, he declared that "I think I'd better take your classes and do the teaching work. You keep on doing what you have to do" (p. 52).

Pursuing his research on α-rays, despite Becquerel's initial claim that they were neither polarized nor refractive, Rutherford proved experimentally that they were indeed deflected by electric and magnetic fields thus inferring that they were particles with a positive charge similar to helium. As early as 1903 he had discovered that, when heated, radioactive substances gave off helium, which convinced him that α-particles were helium ions. Investigating β-rays, the second radioactive material detected by Rutherford, others such as Becquerel, the Curies, and Walter Kaufmann found that they, too, were deflected in electric and magnetic fields, have a negative charge, and an *e/m* ratio similar to cathode rays or electrons. When in 1902 Kaufmann demonstrated that β-rays and cathode rays exhibit very similar properties, from "that time on it was considered settled, the β-rays are electrons." Paul Villard at the École Normale in Paris had discovered a third penetrating emission of radium, called *gamma* (γ) rays, that resembled X-rays.

While working in close collaboration between 1900 and 1903, Rutherford and Soddy made a startling discovery regarding the *transmutation* of radioactive substances, though fearing they might be ridiculed as alchemists, they called the process a "transformation." Their unexpected discovery was that radioactive emissions indicate a change in the *internal structure* of the radioactive substance creating two or more forms of the same element. Soddy introduced the term "isotope" in 1913 to refer to elements that were identical in their chemical properties and position on the Periodic table but differ in their mass and radioactive properties, such Ra^{224}, Ra^{226}, and Ra^{228}. Moreover, when chemically separated from the original radioactive material but left in contact with it, the residue became similarly radioactive and the sum of the curves representing the loss or gain of radioactivity over a prior of time between the two substances remained constant.

Yet, having made these surprising discoveries, physicists were still as baffled by the radioactive process itself as by what produced it, since nothing was yet known about the internal structure of the atom. As Marie Curie described the situation in 1900: "Uranium exhibits no appreciable change of state, no visible chemical transformation; it remains, or so it seems, identical with itself, the source of energy which it emits remains undetectable—and therein lies the profound interest of the phenomenon."[66] Still, the fact that Rutherford and Soddy attributed the radiation to subatomic transmutations was a bold advance. As Pais states, they surmised that "radioactive bodies contain unstable atoms of which a fixed fraction decay per unit time. The rest of the decayed atom is a new radio-element which decays again, and so forth, till finally a stable element is reached" (p. 113). Thus Rutherford and Soddy realized that the emission of energy from the radioactive bodies is caused by the inherent modification of the atoms and therefore does not negate the important law of the conservation of energy.

Though Rutherford received the Nobel Prize in chemistry in 1908 for his outstanding research in radioactivity—such as the discovery of alpha and beta particles and the transmutation of radioactive substances—his major contribution to discovering the internal structure of the atom would not occur until after he had returned to England. In 1921 Soddy also received the Nobel Prize in chemistry for his assistance in discovering radioactive transmutations and the identification of isotopes.

Since what makes an individual a scientist is his or her curiosity and desire to explain ordinary phenomena rather than just accepting them or declaring them miracles, the next discovery marked the beginning of a new inquiry that would further revolutionize the physical sciences. It had been known since household coal furnaces were used that inserting a black poker into the burning coal causes it to change color as the heat increases from red, to violet, to white. It was called "blackbody radiation" because the spectroscopic analysis had established that the various colors corresponded to the different frequencies of the electromagnetic emissions, but without any conception of the internal composition of the atom it was unknown how this occurred.

Since metals are good conductors of electricity and Rutherford had identified the electron as the unit of electrical charge, it was believed that metals somehow contain electrons to explain the flow of electric current. Furthermore, it was hypothesized that the cause of the change in the emitted colors as the temperature increased could be explained if the added heat energy produced an increased oscillation (the still accepted Newtonian explanation) of the electrons thus explaining the increased frequencies. And if the equipartition principle—that any addition of heat was distributed equally among the gas particles increasing their velocities—were true also of blackbody radiation, then since a blackbody absorbs all the incident heat or energy, the distributed increase in energy could account for the increased frequencies of the metallic oscillators.

Thus the spectral distribution and frequency intensities would be indicative of the amount of electromagnetism per unit volume of the blackbody that also could account for the fact that a thermal equilibrium had been reached, as indicated by the specific color. But while there were a number of discovered mathematical laws that contributed to a somewhat more precise correlation between the energy or frequency of the spectral emission and the incident temperature, there still was no explanation as to what accounted for the correlation. Furthermore, the Rayleigh-Jeans Law, governing the path of electromagnetic waves, made the disturbing prediction that the equal distribution of the energy, along with the fact that the higher frequency oscillations radiate their energy more efficiently, would result over time in an increase in the high-frequency radiation tending to infinity. Thus it would be dangerous sitting before a fireplace because of the increasing radiation in the ultraviolet range.

It was called "the ultraviolet catastrophe" owing to the fatal implications and also because it conflicted with the experimentally corroborated fact that the spectral intensity corresponded to the thermal equilibrium as indicated by the specific color and thus did not continue to rise. It was one of "the two dark clouds" (the other being the null results of the Michelson-Morley experiment to be discussed shortly) mentioned by Lord Kelvin that threatened physics at the turn of the century.

This was the situation confronting Max Planck in 1900, who, conservative by nature, introduced his famous formula describing blackbody radiation that issued in a startlingly new turn in physics. Ironically, asserting that "I had always regarded the search for the absolute as the loftiest goal of all scientific activity,"[67] he ended up introducing the greatest *indeterminateness* or *uncertainty* in physics of all time.

His explanation was revolutionary because it indicated that thermodynamic or radiational processes, rather than being continuous as previously assumed, were discrete or discontinuous. Flowing water is an example of a continuous process while a sprinkler or rain drops are examples of discontinuous processes. Planck's task was to find the correct mathematical and physical interpretation of the relation between the energy distribution of the emission density and the entropy distribution of the thermal equilibrium that would eliminate the ultraviolet catastrophe. Years later, when he was asked how he had solved the problem, he replied

> It was an act of desperation. For six years I had struggled with the blackbody theory. I knew the problem was fundamental and I knew the answer. I had to find a theoretical explanation at any cost, except for the inviolability of the two laws of thermodynamics.[68]

The two laws of thermodynamics introduced by R. Clausius in 1885 were first that the energy of the universe is constant and second that the entropy of the universe tends to a maximum. Entropy is the maximum disorder or randomness of the components of a system that is in thermal equilibrium. As Planck admitted, while he was well schooled in thermodynamics he was weak in statistics, so it was Ludwig Boltzmann who first stated the mathematical ratio between entropy and probability as $S = k \log W$, where S stands for entropy, W for the thermodynamic probability, and k is a universal constant. This proved so important that Planck called the k in the formula "Boltzmann's constant."

With Boltzmann's formula Planck was able to calculate the probability of the entropy at its maximum in thermal equilibrium. Then, as Segrè states, he

> divided the energy of an oscillator into small but finite quantities, so that the energy of the oscillators could be written as $E = P\varepsilon$ where P is an integral number. With this hypothesis Planck could calculate the average energy of an oscillator and thus find the blackbody formula. Planck expected that ε could become arbitarily small and that the decomposition of E in finite amounts would only be a calculational device. However, for the results to agree with Wien's thermodynamic law, ε had to be finite and proportional to the frequency of the oscillatore $\varepsilon = hv$, where h is a new universal constant, appropriately called *Planck's constant.* (p. 73)

His equation demonstrated and conformed to the experimental evidence that the energy distributed among the oscillators did not tend to infinitely high frequencies, but to the frequency corresponding to the incident energy, thus eliminating the ultraviolet catastrophe. The physical significance of his discovery is that the electronic oscillators, contrary to classical electrodynamics where the energy could acquire any value in a continuum, according to his formula the energy is restricted to discrete values that are integral multiples of *nhv*, where *n* equals 1, 2, 3, etc. He calculated the value of the constant *h*, known as Planck's constant, to be "6.55×10^{-27} [*erg. sec.*] which today is calculated as $h = 6.6262 \times 10^{-27}$ [*erg. sec*]" (p. 73).

Proving its significance, he derived a number of constants and empirical values from it, one of the strongest indications of a theory's truth. As Segrè further states, even in his first paper "Planck pointed out that from Stefan's law and from Wien's thermodynamical law it is possible to infer the two universal constants *h* [his] and *k*, and from these the charge of the electron, Avogadro's number and more" (p. 73). But despite his success in resolving the problem and receiving the Nobel Prize in physics in 1920 for his achievement, Planck never was reconciled to the conclusion. As he stated:

> I tried immediately to weld the elementary quantum of action somehow in the framework of classical theory. But in the face of all such attempts this constant showed itself to be obdurate. . . . My futile attempts to put the elementary quantum of action into the classical theory continued for a number of years and they cost me a great deal of effort.[69]

Nor was he the only one to be disconcerted by his discovery. Even Einstein had his misgivings about the introduction of quantum mechanics despite his own similar interpretation of the photoelectric effect in terms of *particles of light* called "photons." As he declared in his "Autobiographical Notes," where he discusses the theoretical and experimental background and significance of Planck's discovery:

> Planck got his radiation-formula if he chose his energy-elements ε of the magnitude $\varepsilon = hv$. . . . All of this was quite clear to me shortly after the appearance of Planck's fundamental work; so that, without having a substitute for classical mechanics, I could nevertheless see to what kind of consequences this law of

> temperature-radiation leads. . . . All my attempts, however, to adapt the theoretical foundation of physics to this [new type of knowledge] failed completely. It was as if the ground had been pulled out from under one, with no firm foundation to be seen anywhere, upon which one could have built.[70]

This last quotation points to the year 1905, Einstein's famous *annus mirabilis,* when at age twenty-six he sent six articles to the *Annalen der Physik* (Annals of Physics) between March and December that were promptly published heralding a new era in physics. The first contained his explanation, just described, of the photoelectric effect positing the existence of discrete light quanta or photons, analogous to Planck's explanation of blackbody radiation, that contributed to the emerging revolutionary and contentious science of quantum mechanics. The second, his doctoral thesis, presented his original method for calculating molecular dimensions. The third and sixth pertained to Brownian motion presenting the most convincing evidence then of the existence of molecules. The fourth and fifth introduced his special theory of relativity that dealt with E the electrodynamics of moving bodies from which he derived his famous formula $E = mc^2$, a rival to Newton's F = ma, undoubtedly the two most famous scientific formulas known to humanity.

Although like Planck he was adamantly opposed to the quantum theory throughout his life, his first paper paradoxically was a major contribution to its acceptance, along with explicitly raising the question, perhaps for the first time, of the dual nature of light since photons, as light quanta, complemented the wave theory then accepted. It also was the main reason for awarding him the Nobel Prize in 1921, since his major achievements, the theories of relativity, were unconfirmed at the time.

This is not to imply that his introduction of a strange new particle was universally accepted since it met with considerable skepticism. It was not until Arthur Compton demonstrated in 1923 the scattering of X-rays by electrons, indicating that they followed the same laws of deflection and momentum as colliding particles with the energy hv predicted by Planck and Einstein, that the reality of photons was generally accepted. As usual scientists are generally skeptical of new discoveries until they are confirmed by additional evidence.

It was his third article on Brownian motion, cited more frequently than the others, that was the most successful. Ever since Robert Brown had suspended pollen grains in water in 1827, attributing their observed motion to the impact of the underlying molecules, this had provided the most convincing evidence of their existence. Assuming that the motion of the molecules was due to their size and density, as well as the viscosity and temperature of the water, utilizing the kinetic theory of gases Einstein was able to provide additional evidence of their reality, along with deriving new values for Avogadro's number and for Boltzmann's constant k. When these values were confirmed the uncertainty of the existence of molecules was resolved. As French physicist Jean Perrin stated in 1909:

> I think it is impossible that a mind free from all preconception can reflect upon the extreme diversity of the phenomena which thus converge to the same result without experiencing a strong impression, and I think that it will henceforth be difficult to defend by rational arguments a hostile attitude to molecular hypotheses.[71]

Einstein's fourth and fifth articles dealing with the "electrodynamics of moving bodies" contain his special theory of relativity that led to the refutation of Newton's affirmation of absolute space and time and toward a definite transition to the third conception of reality. Following Newton's publication of the *Principia*, for two centuries absolute space and time had been regarded as necessary ontological frames of the universe to account for absolute motion and rest. Then in the nineteenth century Maxwell's theory of electromagnetism and the reemergence of the wave theory of light led to the hypothesis that Newton's absolute space was filled with a stationary aether to explain the propagation of electromagnetism and light waves.

But not content with merely positing absolute space and time and the aether, between 1881 and 1929 Albert A. Michelson and Edward W. Morley, both independently and in collaboration, devised a number of tests to indirectly confirm the existence of the aether believed to be at rest in absolute space and thus demonstrate the existence also of absolute motion. The most effective of these tests was Michelson's optical method using his interferometer.

Assuming that the earth's annual revolution through the aether would create an aether drift or wind in the opposite direction, one should be able to demonstrate the existence of the aether, and thus of absolute space, by detecting its effect on the velocity of light. Imagine a circular instrument enumerated like the face of a clock, within which there is placed a light source at 9 o'clock and two reflecting mirrors, one at 12 and one at 3 o'clock, and a light detector at 6 o'clock. There also is a mirror set in the center coated in such a way that it either deflects or transmits the light depending on the angle of interaction of the light and the mirror.[72]

According to the experiment, a single beam of light is emitted from the light source at 9 o'clock that splits into two beams, one of which moving horizontally strikes the central mirror and is deflected vertically to the 12 o'clock mirror which then directly reflects it vertically downward *through* the central mirror to the receptor at 6 o'clock. The second beam also moves horizontally but *through* the central mirror to the 3 o'clock mirror that reflects it back to the central mirror that then deflects it to the receptor at 6 o'clock.

As both beams are emitted at the same moment with the same velocity and travel the same distance, but one's trajectory being more vertical and the other's more horizontal, it was predicted that the light beam traveling more in the parallel or vertical direction would, because of the greater aether resistance, be slowed and thus gain an extra 1/25 of a wavelength, which meant that it would return to the detector slightly out of phase. Yet, regardless of how they positioned the interferometer in relation to the to the earth's movement, the two light waves always returned in phase indicating no effect of the aether on their velocity. And so Michelson announced in 1881 that since there was no evidence that an aether drift affected the velocity of light, it can be concluded that it does not exist: "The result of the hypothesis of a stationary aether is thus shown to be incorrect, and the necessary conclusion follows that the hypothesis is erroneous."[73] When Michelson received the Nobel Prize in physics 1907 he was the first American to do so, but it was not for his aether drift experiment, but for his invention of ingenious optical instruments and other experiments.

Considering the significance of the experiment that constituted one of the "two dark clouds" confronting physics, it is surprising that except for Lord Kelvin, Lord Rayleigh, and Lorentz, its startling and contentious results had little impact at the time. Einstein himself declared that it did not play a "decisive role" in his formulation of the special theory of relativity, despite its demonstrating that the velocity of light remained constant with the implication that there was no absolute motion in relation to the aether existing in absolute space, thus disconfirming the existence of the latter but supporting two of his theses presented in his special theory of relativity. Rather than it being the null results of the Michelson-Morely experiments, Einstein indicated what actually influenced him in a talk given in Kyoto, Japan in December 1922.

> I took into consideration Fizeau's experiment and . . . the truth of the Maxwell-Lorentz equation in electrodynamics . . . [that] showed to us [in] the relations of the so-called invariance of the velocity of light that those equations should hold also in the

> moving frame of reference. This invariance of the velocity of light was, however, in conflict with the rule of addition of velocities we knew of well in mechanics. (p. 139; brackets added)

Given the two "considerations" mentioned in this paper, the invariance of the speed of light and the indistinguishability of moving *inertial* systems because the laws of inertial systems are equivalent, Einstein's rejection of Newton's conception of space and time as absolute frames of the universe follows with stunning simplicity. This is because the presuppositions of their calculations are just the converse of each other. Since in Newtonian mechanics the lengths of measuring rods and the rates of clocks are unaffected by their velocities, two systems in relative motion measuring the velocity of another object must find different dimensions. For example, an automobile that has a velocity of 50 miles per hour as measured by a person a rest on the earth will have a velocity of only 10 miles per hour if measured by a driver traveling at 40 miles per hour in the same direction. This is Galileo's well-known "addition of velocities" principle mentioned by Einstein at the end of the previous quotation.

The anomaly is that though this is true of all other measurements, it is not true of light since anyone, regardless of their velocity, will find the velocity of light to be the same. This can be explained, according to Einstein, if the physical measuring instruments, the rods and clocks, are not as Newton assumed and is true of the relatively slight velocities on the earth, unaffected by their velocities. But for those approaching the velocity of light the measuring rods contract and the clocks slow down in ratio to their velocities thereby explaining the identical determination of the velocity of light irrespective of their different velocities. More explicitly, since $v = d/t$, if the measured distance is greater because of the retraction of the measuring rod and the duration longer because of the slowing of the clocks relative to their velocities, then v will increase proportionately to produce the invariant velocity of light.

The formula for determining the degree of retraction and retardation is $\sqrt{(1- v^2/c^2)}$.[74]

Another consequence is that as the system approaches the velocity of light mass increases tending to infinity, which is the reason no physical system can attain that velocity and thus light is a limiting velocity. In his second article on special theory, Einstein declared the equivalence of energy and mass as stated in his famous equation $E = mc^2$ that superseded Newton's famous equation F = ma at greater velocities, a further transition to a different conception of reality due to the limitations of Newtonian science. It also explains the sun's tremendous source of energy and why its gravitation force is so much greater than any other *solar body* because its mass comprises 98 percent of the mass of the solar system.

While Lorentz's and Einstein's transformation equations correlate the spatial, temporal, and mass measurements from a system at rest to one in uniform motion, German mathematician Herman Minkowski in 1904 devised a formula $\sqrt{\{(cT)^2 - R^2\}}$ that gave an identical value for the duration of two events measured by systems in relative uniform motion. As explained by G. J. Whitrow:

> If, according to a particular observer, the difference in time between any two events is T, this associated spatial interval is cT. Then, if R is the space-distance between these two events, Minkowski showed that the difference of the squares of cT and R has the same value for all observers in uniform relative motion. The square root of this quantity is called the *space-time interval* between the two events. Hence although time and three-dimensional space depend on the observer, this new concept of space-time is the same for all observers.[75]

It is this formula, particularly, that gave rise to the conception of the universe as a four-dimensional continuum of events, three of space and one of time, a central feature of Einstein's general theory of relativity.

However, according to this initial interpretation, because both systems are kinematic (devoid of forces) rather than dynamic due to their being inertial, the modifications attributed to the rods, clocks, and mass in the moving system by the one assumed to be at rest are reciprocal, either system being equal as the reference point. Thus in the special theory the effects are merely apparent rather than actual. Similar to the apparent reduction in size or motion of an object seen from a distance, this effect in the special theory is referred to as the "perspective of velocity." It is also true that what has been described applies only to electromagnetic phenomena, not to mechanical or acoustical measurements.

In the latter cases the addition of velocities law holds so that the measurements of distant events, their duration, and whether they are simultaneous are relative to the system of reference. But since the velocity of light is constant it is independent of the motion of its source and of any moving detector. In contrast to velocity, the wavelength and frequency *are* affected, but in such a way that their product, which determines their velocity, remains constant. But since information about any distant event requires some causal transmission such as a light signal, no effect can appear before its cause, contrary to science fiction accounts.

As indicated, in contrast to the special theory of relativity, where the relativistic effects were considered apparent because they involve inertial systems devoid of forces, in the general theory where noninertial velocities are considered, the effects are actual. French physicist Paul Langevin's *voyage au boulet* introduced in 1911, known in English as "the twins paradox," offers a striking illustration. Leaving one twin on the earth with the other boarding a spaceship that accelerates to 1/20,000th less than the speed of light, the twin on the spaceship, owing to the enormous gravitational effect generated by its tremendous acceleration, will have aged two years, while the twin on the earth would have died during the two centuries that had elapsed on the earth.[76] More recently these effects have been confirmed in precise experiments using clocks placed in jet aircrafts circling the earth indicating that the lifetimes of radioactive particles were extended by their greater velocities. Presently, the extreme velocities of the subatomic particles in physics have led to their being incorporated into relativistic quantum field theory.

Not satisfied with the restrictions imposed in the special theory by limiting the relative motions to inertial systems, in the general theory published in 1915 Einstein extended his investigations to nonuniform motions involving forces, but with the same intent of extending his explanations to new dimensions while also attaining greater uniformity and simplicity in the laws of nature. In these endeavors he did not utilize empirical experiments but ingenious thought experiments, such as comparing the effects of being in an elevator falling at the same rate as gravity (so that the free fall cancels the effect of gravity) to being suspended in a gravitational free region of outer space to demonstrate their inertial equivalences; or comparing the effect of being in a gravity free elevator in outer space with being pulled upward with the same accelerating force as an elevator on the earth to show their equivalence.

By these thought experiments he intended to demonstrate the equivalence of gravitational and accelerated motions by merely shifting one's frame of reference, similar to mass and energy being equivalent. In another thought experiment he described falling from the roof of a house and releasing objects of different weights as he fell, concluding that since the falling objects were in the same gravitational field they would seem stationary relative to himself regardless of their weights. As he wrote in an unpublished paper in 1907:

> The gravitational field has only a relative existence in a way similar to the electric field generated by magnetoelectric induction. *Because for an observer falling freely from the roof of a house there exists*—at least in his immediate surroundings—*no*

> *gravitational field.* Indeed, if the observer drops some bodies then these remain relative to him in a state of rest or of uniform motion, independent of their particular chemical or physical nature (in this consideration the air resistance is . . . ignored).
>
> The observer therefore has the right to interpret his state as "at rest."[77]

In the same article he refers to this realization that the "gravitational field has only a relative existence" as "the happiest thought of my life. . . ."

As further vindication of this study showing how science has transformed our conceptions of reality, in 1931 he declared this to be true of Maxwell's field theory:

> Since Maxwell's time, Physical Reality has been thought of as represented by continuous fields . . . not capable of any mechanical interpretation. This change in *the conception of Reality* is the most profound and the most fruitful that physics has experienced since the time of Newton.[78] (italics added)

It was his dream that all the diverse explanations could finally be reduced to one set of laws in a unified field theory to which he dedicated his life—though it eluded him.

While the unification of the electromagnetic force with the strong and weak nuclear forces in the projected grand unified theory (GUT) has partially vindicated his dream, other unifications would prove elusive or unattainable, such as the reduction of gravity to electromagnetism and the elimination of the uncertainties in quantum mechanics. But his emphasis on simplicity, generality, and elegance in the formulation of theories has been enhanced by the recent discoveries of symmetries.

His correct predictions at the end of his 1918 article on the general theory of relativity of the precession of Mercury, the red shift due to the recession of stellar bodies, and especially photographs of the telescopic confirmation by British astronomer Arthur Eddington of the bending of light during a solar eclipse in 1919, and the more recent prediction of black holes from his theory support his claim to have created, at least partially, a new physical reality to replace Newtonian mechanics.

Chapter VII

CONSTRUCTION OF THE ATOM IN THE TWENTIETH CENTURY

Yet the full account of the revolutionary developments in the twentieth century still has not been related, such as the inquiries leading to a more precise conception of the interior structure of the atom. In England one of the first attempts was the "plum pudding" model of J. J. Thomson that consisted of a positively charged mushy sphere on which the negatively charged electrons were embedded, like plums in a pudding, so that their exterior negative charges balanced the positive charge of the mushy interior producing a neutrally charged atom. Among the obvious faults of this model was his attributing the mass of the atom not to the interior substance, but to the exterior electrons, which would prove to be the reverse of the actual structure.

Rutherford, who began his research under Thomson and later would succeed him as Director of the Cavendish Laboratory, having left Cambridge for Montreal, now accepted a position in Manchester, England, were he conducted his own more sophisticated experiments to investigate the interior of the atom. Working in an excellent laboratory and aided by two talented assistants—Hans Geiger, who would invent the Geiger counter for measuring radiation, and Ernest Marsden, who had emigrated from New Zealand to study with his famous compatriot—Rutherford, decided to use his discovered α particles with their positive charge, large mass, and great velocity to probe the interior of the atom.

Instructing his assistants to radiate α particles at thin gold foil and measure the percentage of deflections striking a scintillating screen set at various angles, they found that most of the particles passed directly through the thin foil with a few deflected at *slight* angles by the presumed existence of the interior atoms. He then suggested that Marsden alter the angle of the screen to see if any of the α particles would be deflected at a greater angle and was astonished when Marsden reported that a few had actually been deflected straight backward to the eyepiece, as if they had been repelled by some massive component within the interior of the gold foil. As an indication of his astonishment, Rutherford described his reaction as "quite the most incredible event that has ever happened to me in my life. It was almost as incredible as if you fired a 15-inch shell at a piece of tissue paper and it came back and hit you."[79]

Undeterred by his astonishment, he began experiments to find a more precise explanation of the cause. Based on the measurements of the percentages of deflections at various angles, he devised a formula for measuring the angles of deflection, the velocity of the particles, and their charges. He also conceived of the nucleus as consisting of particles of a certain mass, along with a positive charge calculated by his formula indicating the atomic number of the element. Accordingly, he defined two of the properties of the nucleus, the atomic mass and the atomic number, that were independently confirmed by J. J. Thomson and H. G. J. Moseley based on X-ray emissions from the atom.

Apparently, from these X-ray emissions it was inferred that the atomic nucleus was surrounded by the negatively charged electrons, which are relatively massless, about 9×10^{-28} g, or 1,836 times less than that of the proton, but whose shells occupy most of the volume of the atom confirming an earlier conjecture by Jean-Baptiste Perrin in 1901 that the structure of the atom might resemble the solar system:

> Each atom might consist . . . of one or more positive suns . . . and small negative planets. . . . If the atom is quite heavy, the corpuscle farthest from the centre . . . will be poorly held by the electrical attraction . . . . The slightest cause will detach it; the formation of cathode rays [electrons] will become so easy that [such] matter will appear spontaneously "radioactive. . . ."[80]

Indicative of how much progress was being made in explaining the structure of the atom Segrè states that the

> new science of X-ray spectroscopy not only allows the study of deep electron shells and elementary chemical analysis on an unprecedented level of sensitivity and certainty: it also opens the way to the exploration of crystalline lattices and, more generally the architecture of solids and of molecules.[81]

It was decided that electrons in the outer shell cause the visible spectra, while the electrons in the inward shell are the source of the X-ray spectra. Apparently, it was Moseley's and Thomson's induced X-ray spectra that provided the evidence for the electron shells and Moseley who determined that the electron shells are related to the nuclear charge, thus providing independent evidence of the atomic number.

Becoming convinced that the experimental evidence supported his calculations, Rutherford presented his results first to the Manchester Literary and Philosophical Society in March 1911—the same society to which Dalton had submitted his atomic theory—and then sent a more detailed account in May to the *Philosophical Magazine* followed by another article entitled "The Structure of the Atom" in February 1914. Though he was unable to explain the exact causes of either atomic stability or radioactive instability, his conception of the composition of the nucleus and structure of the electron orbits was sufficient to enable physicists to formulate a clearer notational designation of the nuclear components and properties of the atom. For example, depicting the *charge* as plus or minus e and the *number* of the charged units as Z, then $+Ze$ stood for the total charge of the nucleus with $-Ze$ representing the total charge of the number of electrons in a particular atom.

Thus if there is an equal number of $+Ze$s and $-Ze$s, the charge of the atom is neutral, while ionization consists of the loss or gain of electrons and radioactive transmutations as a change in the nuclear number due to the emission of α, β, or γ rays. As Ne'eman and Kirsh state:

> While emitting an alpha particle, the nucleus loses electric charge of $+2e$ and a mass of about 4 amu. The process, which is also called alpha decay, or disintegration, lowers the atomic number Z by 2, and the mass number A by 4. The equation representing the alpha decay of uranium 238, for example, is: ${}_{92}U^{238} \rightarrow {}_{90}Th^{234} + {}_{2}He^{4}$.[82]

Since the number of $-Ze$ represents the number of electrons in the atom, which accounts for the chemical properties, it also indicates its place in Mendeleev's Periodic Table. Since in neutral atoms the number of $+Ze$s equals the number of $-Ze$s, isotopes are atoms with identical chemical properties but different atomic weights. Yet while the chemical properties were attributed to the electrons, it still was not known what accounted for the *nuclear* numbers and weights. In 1919 Rutherford had discovered that when an a particle interacts with a hydrogen atom a hydrogen nuclear particle is ejected, but did this mean that atomic nuclei were all hydrogen nuclei?

After a number of experiments probing the nuclei of other atoms produced the ejection of the same entity, physicists decided they had discovered the first nuclear particle, naming it "proton," after the Greek word *protos* meaning "first." But they were still puzzled by the fact that the number of the nuclear particles of an atom did not match its mass or atomic weight. However, when James Chadwick, at the Cavendish Laboratory, probed the nuclei of lighter elements, such as beryllium, he discovered that a new particle was ejected that was quite massive but neutral in charge. Determining the mass of the particle to be nearly that of the proton and finding that it had a neutral charge, for obvious reasons he named it a "neutron" and received the Nobel Prize in 1935 for his discovery. Thus the discovery of subatomic particles was resolving many problems at a stroke, as Ne'eman and Kirsh affirm.

> The discovery of the neutron is a classical example of the way in which the addition of a new building block clarifies as if by magic many previously inexplicable facts. For example, it became clear that the mass number A is just the total number of protons and neutrons in the nucleus. The fact that the atomic mass is always quite close to an integral number of amu found simple explanation: the masses of both the proton and the neutron are close to 1 amu. Different isotopes of an element are atoms the nuclei of which have the same number of protons but not the same number of neutrons. (p. 18)

Following his success in detecting the proton, deciding to eject α particles into the air Rutherford succeeded in producing nuclear disintegration. In an article titled "Collision of Alpha Particles with Light Atoms," published in the June 1919 issue of the *Philosophical Magazine*, he describes in the fourth part "An Anomalous Effect in Nitrogen" in which the

> nitrogen atom is disintegrated under the intense forces developed in a close collision with a swift alpha particle, and that the hydrogen atom which is liberated formed a constituent part of the nitrogen nucleus. . . . The results . . . suggest that if α particles—or similar projectiles—of still greater energy were available for experiment, we might expect to break down the nucleus structure of many of the lighter atoms.[83]

Yet the model was mainly due to the great Danish physicist from Copenhagen, Niels Bohr (1885–1962), whom most physicists consider, along with Einstein and Rutherford, one of the three greatest physicists of the new era. Along with Einstein, he was a dominant influence on the development of theoretical physics in the first half of the twentieth century due to his own contributions and those of his famous Institute of Theoretical Physics in Copenhagen at which all the famous physicists of the era, such as Hendrik "Hans" Kramer, Wolfgang Pauli, Werner Heisenberg, Erwin Schrödinger, and Paul Dirac, etc., at one time attended.

The Cavendish Laboratory under the direction of Thomson being the most outstanding research center in physics, Bohr decided to do his postgraduate study there under the supervision of Thomson. However, having studied Thomson's model of the atom carefully and found some faults in it, during their very first meeting after his arrival in the fall of 1911, he had the temerity to point them out to the famed physicist. Apparently not being accustomed to being corrected by a twenty-six-year-old graduate student who barely spoke English, Thomson did not appreciate the criticisms so the relationship, though not hostile, was not friendly.

Fortunately, however, in November Bohr met Rutherford in Manchester and then heard him lecture in Cambridge on his developing theory of the atom. These encounters led to his amicable departure from the Cavendish to study in Rutherford's laboratory in Manchester where his association with Rutherford flourished to the extend that he considered him a kind of surrogate for his father who had died earlier. They remained close friends and admirers until Rutherford's death in 1937.

Arriving in Manchester in March 1912, Bohr took courses in radioactivity and then, at the suggestion of Rutherford, began his own experiments investigating the nucleus with α particles. Realizing, however, that these nuclear experiments were not conducive to his primary theoretical interest in locating the electrons in the atom he decided to forego them. At the suggestion of two other research associates, Gorge von Hevesy, who was to become famous for his ingenious research in radioactivity, and Charles Darwin, the grandson of the famous evolutionist, who was experimenting on the effects of α particles on electrons, Bohr also began to investigate the impact of α particles on clusters of electrons to determine their placement within the atom.

In the summer of 1912 he prepared a draft article "On the Constitution of Atoms and Molecules," known as the "Rutherford Memorandum" (which he showed to Rutherford), containing his criticisms of Thomson's conception of the atom. First among the criticisms was the static placement of the electrons in the plum pudding model that Bohr thought untenable, yet, according to classical electrodynamics, if they revolved around the nucleus they would continue to radiate energy and eventually spiral into the nucleus, which also was unacceptable. Second, they could not rotate in the same shell because on classical principles that, too, would prove unstable. Third, on Thomson's model the atom's radius was not determined by the rotating electrons, but by the mushy positive interior, another apparent misconception.

Concluding that it would be "hopeless" to try resolving these problems in terms of classical mechanics, Bohr decided to incorporate Planck and Einstein's quantification of radiation in his calculations. Like others at the time, initially he conceived the electrons as "atomic vibrators" that radiated according to Planck's quantum of action as Bohr wrote in one of his famous articles published in 1913:

> According to Planck's theory of radiation . . . the smallest quantity of energy which can be radiated out from an atomic vibrator is equal to *vk*, where *v* is the number of vibrations per second and *k* (we now called it *h*) is Planck's constant. *Thus did the quantum theory enter the interior of the atom for the first time.*[84]

But the difficulty was acquiring the empirical data that would contribute to solving the problem. During the short time Bohr was at the Cavendish Laboratory he probably met John Nicholson, a gifted young researcher who in 1911 "associated spectral lines with various modes of vibration of electrons around their equilibrium orbits in the field of a central charge."[85] In another article he proposed that the angular momentum of a revolving electron could be calculated from the mass, velocity, and the radius of its orbit. As Pais states: "If, therefore, the constant *h* of Planck has . . . an atomic significance, it may mean that the angular momentum of an atom can only rise or fall by discrete amounts when electrons leave or return" (p. 199).

But it was a suggestion of a student friend that led Bohr to consider the possible effect of electron orbits (rather than atomic oscillations) on the spectral emissions that produce the frequencies, especially in connection with Johann Balmer's radiational formula introduced in 1885. The latter depicts the related mathematical frequencies of light emitted by hydrogen into a small number of series that provide the crucial insight. As Bohr recalled many years later: "As soon as I saw Balmer's formula the whole thing was immediately clear to me. . . ."[86]

What Bohr immediately realized was that Planck's constant of action *h* at 6.62×10^{-27} erg-seconds was the key to solving the problem of the orbits of electrons if he included it in his formula. Because *h* has a fixed value, by inserting it meant that the electron's orbits could only take on specified values that were multiples of Planck's constant. In addition, this indicated that any change from one orbit to another would require an exchange of energy: the absorption of energy if the electron jumped to a higher orbit or a discharge of energy if the electron dropped to a lower orbit. Furthermore, he inferred that since the exchange of energy came in a predetermined amount, it must be a form of electromagnetic radiation and also that $E = hv$ meant the radiation involved definite frequencies since v stands for frequencies.

Thus Bohr's solar model of the atom consisted of electrons surrounding the nucleus in a series of orbits called "stationary states" whose *locations* were dictated by Planck's constant but whose *orbital trajectories* were determined by their angular momentum and energy or frequency. The emissions originated from a spaceless "jump" of the atomic electrons from one orbital or energy level to another. This also showed that not all transmutations originate in the nucleus as previously believed.

But there remained the original problem of why, according to classical physics, the electron at the lowest orbital state was not drawn into the nucleus by the electrostatic force eventually overcoming the energy of its angular momentum. This was solved by Bohr simply declaring that the "ground state is stable," which Pais describes as "one of the most audacious postulates ever seen in physics."[87]

Bohr's explanation was also "audacious" because it incorporated classical and quantum mechanics in an unorthodox way: he described the *energy* or *frequencies* of the particles' angular momentum in terms of *classical mechanics,* while the *optical frequencies* of the ejected photons were described wholly in terms of *quantum mechanics.* This use of two conflicting mechanistic explanations, although related to different kinds of frequencies, is the first evidence of the two principles that would guide Bohr's interpretation of quantum mechanics: the "correspondence principle" and the "principle of complementarity."

Where the magnitudes of the frequencies of the *electrons* in the successive orbits resemble classical magnitudes the calculations tend to "correspond," but where the magnitude of the *optical* frequencies of the ejected photons is much greater they tend to be "complimentary." While unorthodox, the solution is similar in physics where relatively slow velocities are computed in terms of Newtonian mechanics and the greater optical velocities are computed by Einstein's relativity theory.

Despite his model being largely theoretical, the derivations Bohr was able to infer from them were amazing, such as Balmer's formula, Rydberg's constant, deducing the radius of the bound state of the stable hydrogen atom (called the "Bohr's radius"), and especially demonstrating that a series of stellar spectral lines that had been attributed to hydrogen were in fact the spectral lines of ionized helium.

In addition to these computational achievements, he was able to devise a number of new theoretical explanations of spectral emissions using such theories as blackbody radiation, the photoelectric effect, nuclear radioactivity, electron radiation, and spectroscopy. Also, by dividing the electron orbits into outer and inner shells he could attribute visible spectra to the outer electron shells while X-rays he explained were due to an electron being ejected from an inner shell. Beta rays, later identified as electrons, he correctly attributed to radioactive decay in the nucleus.

Although the above is just a cursory summary it does convey something of the magnitude of his accomplishments. Similar to Einstein's five publications in 1905, at the early age of twenty-five, that revolutionized Newtonian mechanics, Bohr's three publications in 1913 when he was twenty-eight years old on the orbital structure and function of electrons revolutionized atomic physics. Pais summarizes Bohr's accomplishments:

> The very existence of line (and band) spectra suggests, he noted, that electrons move in discrete stationary orbits inside atoms and molecules. Spectra (including X-ray spectra) arise because of quantum jumps between these states. . . . The quantitative confirmation of these ideas by his treatment of hydrogen and ionized helium mark a turning point in the physics of the twentieth century and the high point in Bohr's creative career. The insistence on the role of the outermost ring of electrons as the seat of most chemical properties of the elements, in particular their valencies, constitutes the first step toward quantum

> chemistry. The sharp distinction between atomic/molecular and nuclear physics begins with his realization that β-rays emanate from the nucleus.[88]

The response to these innovations, called the "Copenhagen interpretation," were diverse with Rutherford, Arnold Sommerfeld, and Moseley favorable, while Otto Stern and Max von Laue declared, after reading Bohr's 1913 article that, "if by chance it should prove correct, they would quit physics" (though they later changed their minds). It was Einstein who paid the most eloquent tribute becoming a close friend and admirer of Bohr despite his strong aversion later to quantum mechanics.

> That this insecure and contradictory foundation was sufficient to enable a man of Bohr's unique instinct and tact to discover the major laws of the spectral lines and the electron shells of the atoms together with their significance for chemistry appeared to me like a miracle—and appears to me as a miracle even today. This is the highest form of musicality in the sphere of thought.[89]

But Bohr, aware of its ad hoc nature was somewhat uncertain of his theory of the hydrogen spectra, yet because of the general agreement of his solar model of the atom with the experimental evidence and consistency with other quantitative derivations, he accepted it as a provisional theory despite knowing that, due to the uncertain nature of the measurements, it lacked a definitive internal structure and firm independent reality. Still, as Pais indicates, his achievements were exceptional.

> As the year 1913 began, almost unanimous consensus had been reached, after much struggle, that atoms are real. Even before that year it had become evident that atoms have substructure, but no one yet knew by what rules their parts moved. During that year, Bohr, fully conscious that these motions could not possibly

> be described in terms of classical physics, but that it nevertheless was essential to establish a link between classical and quantum physics, gave the first firm and lasing direction toward an understanding of atomic structure and atomic dynamics. In that sense he may be considered the father of the atom.[90]

Although at the time Bohr's solar model of the atom was mainly conjectural, its authenticity was reinforced by the supporting discoveries it engendered, to which we now turn. Among the things he could not explain was the reason that electrons in the hydrogen atom were restricted to their particular orbits and why their angular momentum was an integral multiple of Planck's constant h, or depict the atomic spectra of atoms with an electronic structure more complex than hydrogen. This awaited the discoveries of the "new quantum theory," in contrast to Bohr's "older quantum theory," begun in 1923 as a result of the successive contributions mainly of A. H. Compton, Louis de Broglie, Werner Heisenberg, Paul Dirac, and Erwin Schrödinger, along with Wolfgang Pauli, Pascual Jordan, and Max Born.

One of the first to begin the transition was Prince Louis de Broglie who, assisting his older brother Duke Maurice in his experiments on the dual properties of light, presented in 1923 his doctoral dissertation to the French Academy of Sciences on "The Connection between Waves and Particles." Recall that to explain blackbody radiation Planck had introduced in 1900 the concept of the "quantum of energy" to designate the discontinuous or discrete *exchange of energy* caused by the light waves striking the oscillating electrons within the metal to produce the ejected electrons, but he had restricted it solely to the *interaction* occurring *within the metal.* In contrast, in 1905 Einstein explained the photoelectric effect as not due to *light waves* striking the metallic surface but to a stream of *light particles* colliding with the electrons within the metal, which light particles were later named "photons."

Having no mass, in order to eject the electrons from within the metal the light particles had to possess a certain quantity of energy that Einstein defined as hf (Planck's constant h times the light frequency f) per particle. So, though light produced diffraction patterns characteristic of waves when injected through a tiny aperture, it also displayed the properties of particles when reflected on metals causing an interaction with the inner vibrating electrons. This duality of contrary properties introduced one of the two main paradoxes of quantum mechanics, the other being the uncertainty of the measured properties.

The duality was further evidenced when in 1922 the American physicist Arthur Compton discovered that when paraffin was irradiated with X-rays a portion of the emerging waves had longer wavelengths than the entering ones. Compton believed the effect was not explainable by the usual wave theory of electromagnetic radiation but could be explained if light consisted of quanta of energy some of which was transmitted to the electrons in the paraffin due to the collisions. So even if the light quanta do not have mass, they do have momentum owing to the energy.

His experiment showed that "a photon of electromagnetic radiation of wavelength λ carries momentum whose value is $p = h/\lambda$, where h is Planck's constant, λ the wave length, and p the momentum. Compton explained that in the collision with an electron the photon loses momentum, and therefore its wavelength increases by the scattering."[91] Thus the explanation of blackbody radiation, the photoelectric effect, and Compton's experiment reinforced the validity of the dualism of light despite the paradoxical contradiction of properties.

Intrigued by light having particle properties, de Broglie decided to investigate as his dissertation subject whether the converse was true, that particles could have wave properties. Ne'eman and Kirsh summarize his revolutionary theory as follows.

> Every moving particle has an associated wave of definite wavelength and frequency determined by the mass and velocity of

> the particle. The laws of motion of small particles cannot be understood unless the wave nature of the particles is taken into account, just as the photoelectric effect and black body radiation cannot be understood without resort to the particle properties of light.
>
> The mathematics of this model was simple. De Broglie assumed that the equations $E = hf$ and $p = h/\lambda$ were valid for material particles as for photons. Thus, the wavelength, λ, of the particle is given by $\lambda = h/p$ where p is its momentum. The faster the particle moves, the shorter is its wavelength. (pp. 39–40)

De Broglie's dualistic theory contributed to Bohr's orbital model of the atom by explaining that if electrons possessed wave properties, the circumference of the orbital circle would have to be an integral number of wavelengths, otherwise they would interfere and destroy. He also calculated the angular momentum of the circular orbits of the electrons that could be observed in a cloud chamber. Yet other questions pertaining to Bohr's model were still unexplained: why the electrons in the hydrogen atom were restricted to their particular orbits? what accounts for their orbital jumps? and what were the atomic spectra of atoms that had an electronic composition more complex than hydrogen?

Furthermore, there was the paradox that while the path of an electron is determined by Newton's mechanistic laws its stability within the atom required the laws of quantum theory. Since the latter laws did not conform with Newtonian laws a new set of laws and theoretical framework had to be formulated, named "quantum mechanics," to describe this new system of subatomic particles. The initially proposed explanations were the "matrix mechanics" of Werner Heisenberg (1901–1976) and the contrasting "wave mechanics" of Erwin Schrödinger (1887–1961).

If one would like to know of a readable account of the turmoil facing young physicists in the early twentieth century owing to Bohr's introduction of the solar model of the atom, the dualism between the wave and the particle properties of electrons, and the necessity of using both Newtonian and quantum laws, I know of no more fascinating book than Heisenberg's biographical account, *Physics and Beyond: Encounters and Conversations* (1971), recounting his intellectual development and outstanding contributions to the current revolutionary developments in physics quoted earlier.

Owing to his decisions to study atomic physics, enroll at the University of Munich, and attend the lectures of Arnold Sommerfeld from 1920 to 1922, he was privileged to have been invited by Sommerfeld to hear the lectures by Bohr, then age thirty-seven, in the summer of 1922 at what has come to be known as the "Göttingen Bohr Festival" at Göttingen's famous school of mathematics. Having been introduced to Bohr by Sommerfeld and then, during the question period following a lecture titled "Advanced Objections," to Hendrik "Hans" Kramer's speculations on quantum mechanics discussed by Bohr during the lecture, Bohr was so impressed by Heisenberg's comments that at the end of the discussion period he asked Heisenberg to join him on a walk that afternoon over the Hain Mountain. As Heisenberg wrote: "This walk was to have profound repercussions on my scientific career, or perhaps it is more correct to say that my real scientific career only began that afternoon."[92]

During their walk Bohr began the conversation by indicating what initially had been his main concern in physics, saying that it was not what one might think, the inner structure of atoms, but the "stability of nature," an insight also as to how a great scientist thinks. Giving examples to show that classical Newtonian science took for granted the normal stability and uniformity of nature in formulating the laws and mathematical equations that describe the causal connections and inner structure of the Newtonian corpuscular-mechanistic worldview, Bohr's basic motivation was not merely to describe their stability and uniformity *but to explain what accounts for it.*

As quoted by Heisenberg, Bohr stated:

> My starting point was not at all the idea that an atom is a small-scale planetary system and as such governed by the laws of astronomy. I never took things as literally as that. My starting point was rather the stability of matter, a pure miracle when considered from the standpoint of classical physics. (p. 39)

He then explained what he meant by the stability, persistence, and recurrence of the physical and chemical properties of the elements in terms of their combining in definite proportions to constitute the molecular structures of certain substances while retaining their original properties when decomposed, which forms the basis of the classifications in Mendeleev's Periodic Table. Still, the connection between the stability in nature and Bohr's fascination with atomic structure is revealed in a statement he made, again quoted by Heisenberg:

> The existence of uniform substances, of solid bodies, depends on the stability of atoms; that is . . . quite inexplicable in terms of the basic principle of Newtonian physics, according to which all effects have precisely determined causes, and according to which the present state of a phenomenon or process is fully determined by the one that immediately preceded it. This fact used to disturb me a great deal when I first began to look into atomic physics. (p. 39)

This also reveals the radical distinction between the scientific orientation of Bohr and Einstein, despite their close friendship and enduring admiration for each other, the latter searching for a formula that would describe a four-dimensional space-time field that would precisely unify all the laws and structure of the universe while the former was trying to determine the nature of the subatomic structures that produce the underlying probabilistic causal interactions in nature. Accordingly, Bohr describes all the developments in the past few decades that led to the realization of the limitations of classical physics and the need to replace or supplement it by an understanding, however tenuous, of the deeper subatomic realm (pp. 39-40).

Bohr then presents his position that seems to have had a lasting effect on Heisenberg. This position affirms that since classical physics deals with our familiar macroscopic world it can provide visual or verbal representations of that world while subatomic or quantum physics investigates a domain that is so unique that it does not allow such ordinary descriptions. As Bohr clearly states in a quotation by Heisenberg:

> We know from the stability of matter that Newtonian physics does not apply to the interior of the atom; at best it can occasionally offer us a guideline. It follows that there can be no descriptive account of the structure of the atom; all such accounts must necessarily be based on classical concepts which, as we saw, no longer apply. You see that anyone trying to develop such as theory is really trying the impossible. For we intend to say something about the structure of the atom but lack a language in which we can make ourselves understood. (p. 40)

In response to these skeptical reflections, Heisenberg bluntly asks Bohr what was the point of his introducing "all those atomic models" in his lectures and what did he expect to show by them, to which Bohr replies: "These models have been deduced, or if you prefer guessed, from experiments, not from theoretical calculations. I hope that they describe the structure of the atoms as well, but *only* as well, as is possible in the descriptive language of classical physics" (p. 41).

Heisenberg next asks how can we ever expect to understand the nature of atoms if a description of their inner structure is not clearly defined? Hesitating momentarily, Bohr replies, "I think we may yet be able to do so. But in the process we may have to learn what the word 'understanding' really means" (p. 41). His answer foretells the eventual "uncertainty" in the conception of the interior structure of the atom as well as those inherent in quantum mechanics in general.

Impressed by the conversation with Heisenberg, Bohr invited him to Copenhagen to attend Bohr's Institute, but finding Bohr's attempt to construct the inner structure of atoms on the experimental evidence unappealing, Heisenberg decided instead to go to Göttingen in the fall of 1924 to study with Max Born whose approach to atomic physics was based more on the mathematical calculations. Owing to his father being a professor of Greek at the University of Munich, Heisenberg had learned to read the Greek classics in the original when he was just sixteen years old. Thus, like Kepler, he became attracted to atomic physics by reading Plato's *Timaeus*, which equated the four basic elements of fire, earth, air, and water, along with the cosmos, to the five Pythagorean geometric solids. Despite his realization that the endeavor seemed to be "wild speculation," Heisenberg "was enthralled by the idea that the smallest particles of matter must reduce to some mathematical form (p. 8).

Suffering from a severe attack of hay fever in Munich the following summer, Heisenberg sought shelter in the small island of Heligoland on the North Sea where the sea breeze dispersed any pollen-laden air. It was there he began writing a paper on "electron states" based entirely on the measured light frequencies absorbed or emitted by the atom that was published in *Zeitschrift für Physik* (Writing on Physics) in September 1925. As he indicated, despite rejecting any attempt to visualize the interior structure of the atom, like Plato he seemed to find in the numbers a kind of abstract mathematical reflection of its interior.

> Within a few days . . . it had become clear to me what precisely had to take the place of the Bohr-Sommerfeld quantum conditions in an atomic physics working with none but observable [or measurable] magnitudes. . . . The energy principle had held for all the terms, and I could no longer doubt the mathematical

> consistency and coherence of the kind of quantum mechanics to which my calculations pointed. At first, I was deeply alarmed. I had the feeling that, through the surface of atomic phenomena, I was looking at a strangely beautiful interior, and felt almost giddy at the thought that I now had to probe this wealth of mathematical structures nature had so generously spread out before me. (p. 61; brackets added)

The paper did contain peculiar mathematical functions when the matrices were added, subtracted, and especially when multiplied, unlike those of traditional mathematics, that he was unable to explain. But given the fact that such famous mathematicians as David Hilbert, Richard Courant, and Max Born were among his colleagues at Göttingen University, when shown to Born the latter recognized the incongruities, especially the violation of what is known as the commutative law. The law states that when multiplying two numbers the order of the numbers is irrelevant: $9 \times 8 = 8 \times 9$, but Heisenberg did not find this to be true of his calculations.

When shown the paper Born was not perplexed because he recognized it as an example of matrix mechanics. As authors Crease and Mann state:

> Born was one of the few physicists in Europe—perhaps the only one—with a good knowledge of matrix mathematics. He realized that Heisenberg's quantum-theoretical series were nothing more, nothing less, than awkward manipulations of frequency matrices. . . . Born was delighted. Rewriting Heisenberg's equations as matrices led to a whole new world of applications he could explore.[93]

Yet it still required some interpretation as the continuing quotation indicates:

> The first thing he figured out was that the matrix *q* for position and the matrix *p* for momentum are noncommutative in a very

> special way: that is, *pq* is not only different from *qp*, but the difference between *pq* and *qp* is always the same amount, no matter what *p* or *q* you chose. Mathematically he wrote this $pq - qp = \hbar/i$, where $\hbar$, as usual, is Planck's constant divided by twice pi, and *i* is the special symbol mathematicians use for the square root of minus one. (p. 50)

But still not satisfied Born, with the aid of Pascual Jordan (who wrote most of the article because Born had suffered a nervous collapse), published a paper also in the *Zeitschrift für Physik* that described the basic mathematics of what later became known as matrix mechanics. As described by Ne'eman and Kirsh:

> They arranged the measurable quantities in square arrays of numbers (such arrays are called "matrices") and by defining mathematical operations between these matrices, they created a consistent quantum mechanical theory. This version of quantum mechanics, which has been known as matrix mechanics, succeeded in explaining certain experimental facts in atomic physics and even predicted unknown phenomena which were verified later experimentally.[94]

For their contributions Heisenberg won the Nobel Prize in 1932, when he was only thirty-one years old; Dirac, together with Schrödinger, in 1933; and Born in 1954. Ne'eman and Kirsh add that the "number of important contributions" Heisenberg "made to physics exceeded that of any other physicist in the twentieth century, except Einstein" (p. 44). I would state, and Bohr.

As an indication of how intriguing the mathematics of quantum mechanics was in those early days, Paul Dirac, a close friend of Heisenberg with whom he had discussed atomic theory in their student days, and who collaborated with Born in writing several articles, also favored a mathematical construction of the interior of the atom rather than Bohr's solar model. After receiving a copy of Heisenberg's paper and studying it for little over a week, he "sat down and wrote an alternative formulation, which presented quantum mechanics as a coherent axiomatic theory" (p. 44).

In my discussion I may have quoted statements that gave the impression that Heisenberg remained very satisfied with his introduction of matrices in place of Bohr's pictorial electron orbits, but later I read a statement by him quoted by Crease and Mann that apparently shows his misgivings in working with a purely numerical system such as matrices despite its earning him the Nobel Prize. It was written to Pauli after Heisenberg wrote his famous article in collaboration with Jordan and Born published in the *Zeitschrift für Physik.*

> *I've taken a lot of trouble to make the work physical, and I'm relatively content with it. But I'm still pretty unhappy with the theory as a whole and I was delighted that you were completely on my side about [the relative roles] of mathematics and physics. Here I'm in an environment that thinks and feels exactly the opposite way, and I don't know whether I'm just too stupid to understand the mathematics. Göttingen is divided into two camps: one, which speaks, like [the prominent mathematician David Hilbert] (and [another mathematical physicist, Herman Weyl], in a letter to Jordan), of the great success that will follow the development of matrix calculations in physics; the other, which, like [physicist James Franck], maintains that the matrices will never be understood. I'm always annoyed when I hear the theory going by the name of matrix physics. For awhile, I intended to strike the word "matrix" completely out of the paper and replace it with another [term] — "quantum-theoretical quantity," for example.*[95] (italics, brackets, and parentheses are in the original)

How different his present dissatisfaction with matrices is from his youthful intention of replacing physical objects with mathematical symbols.

But as might be expected, not all the physicists at the time were disdainful of visual model building, the Viennese-born Erwin Schrödinger (1887–1961) being an outstanding example.

Attracted to de Broglie's discovery that particles also have wave properties that are physical and observable, he was the physicist Einstein consulted when asked by the committee examining de Broglie's doctoral dissertation of its credibility, but also being undecided Einstein did not give his approval until Schrödinger assured him that the thesis had merit.

Schrödinger, influenced by Einstein's belief that fields should replace material particles as the fundamental reality along with de Broglie's discovery that particles have wave properties, decided to see if a complete wave theory could be formulated to replace Bohr's electronic solar model. De Broglie had shown that although mass and momentum were considered properties of particles, they also could be depicted as functions of waves. Such equivalences had previously been discovered: Einstein showing the equivalence of mass and energy ($E = mc^2$) and Planck that energy could be equated with frequency ($e = hv$). Since waves have energy and energy has mass, it was possible that material particles could be depicted as waves as well as particles.

Starting with the classical wave equation that describes the spatial properties of electromagnetic waves, Schrödinger began investigating whether a wave equation could be found to describe the wave properties of subatomic particles, one that would supplement the equations of Newtonian mechanics. In four papers he presented his new theory of "wave mechanics" that appeared in *Annalen der Physik* (Annals of Physics) from January to April 1926 with the title "Quantization as an Eigenvalue Problem."

Among his publications he proposed that waves should be considered the basic reality and introduced what is now known as the "celebrated" Schrödinger's wave equation $i\hbar \frac{\partial}{\partial t} | \Psi \rangle = H | \Psi \rangle$ that contains the famous scalar wave function Ψ (psi), along with the Hamiltonian H, "which is simply the observable corresponding to the energy of the system under consideration."[96] As explained by Pais, in replacing particles with waves he

> suggested that waves are the basic reality, particles are only derivative things. In support of this monistic view he considered a wave packet made up out of linear harmonic oscillator wavefunctions . . . a superposition of eigenfunctions so chosen that at a given time the packet looks like a blob localized in a more or less small region. . . . He examined what happened to his packet in the course of time and found: "Our wave packet holds *permanently together*, does *not* expand over an ever greater domain in the course of time." This result led him to anticipate that a particle is nothing more nor less than a very confined packet of waves, and that, therefore, wave mechanics would turn out to be a branch of classical physics, a new branch, to be sure, yet as classical as the theory of vibrating strings or drums or balls.[97]

According to the equation, between measurements the state vector known as the "wave function" moves in an undisturbed, regular way as described in classical physics until it is measured, which then causes it to collapse into an eigenvalue or single value that produces the observation. Thus its state is uncertain until measured. Owing to its being an extension of classical wave theory, formulated in a mathematics more familiar than matrix mechanics, and providing a visualizable explanation, it was acclaimed by most physicists. As described by Crease and Mann,

> the mathematics Schrödinger used was much easier for physicists to understand. . . . If it was hard to imagine how a solid object like an atom could really be made out of waves—what was making the waves?—many physicists had confidence that Schrödinger, a clever fellow, would figure out the answer.[98]

Moreover, as also explained by Crease and Mann, Schrödinger even proposed an explanation as to how particles could be considered as waves.

> A particle was in reality nothing but "a group of waves of relatively small dimensions in every direction," that is, a sort of tiny clump of waves, its behavior governed by wave interactions. Ordinarily, the bundle of waves was small enough that one could think of it as a dot, a point, a particle in the old sense. But in the microworld, Schrödinger argued, this approximation broke down. There it became useless to talk about particles. At very small distances, "we *must* proceed strictly according to the wave theory, that is, we must proceed from the *wave equation*, and not from the fundamental equation of mechanics, in order to include all possible processes." (p. 56)

Unfortunately, these expectations turned out to be illusory. Instead of determinate or distinct portrayals of the electron states of the atom, the solutions to Schrödinger's wave equation produced small cloudlike images reminding one of Rorschach ink blots. Yet even Born, who contributed to the article creating matrix mechanics, after reading Schrödinger's first paper wrote that he was drawn to the traditional aspects of Schrödinger's wave mechanics, a view that angered Heisenberg.

But then, surprisingly, on April 12, 1926, after a very careful perusal of both matrix mechanics and Schrödinger's wave mechanics,

> Pauli sent a lengthy letter to Jordan in which he proved that the two approaches were identical [or more accurately stated mathematically equivalent]. Schrödinger himself proved the same thing, a little less completely, a month later. . . . In the equivalence paper, Schrödinger mentions *pro forma*, that it was really impossible to decide between the two theories—and then went on to argue fiercely the merits of wave mechanics. (p. 57; brackets added)

Yet despite his attraction to the more traditional approach of Schrödinger's wave mechanics compared to matrix mechanics, during his investigations Max Born made a discovery described in two papers titled (in translation), "Quantum Mechanics of Collision Phenomena," published in the *Zeitschrift für Physik* in June and July of 1926 that challenged Schrödinger's claim that wave mechanics, based on measurements of actual waves, was closer to classical physics than matrix mechanics, which dealt only with abstract numerical matrices. The June paper discovered an indeterminacy or uncertainty in Schrödinger's method of determining the position of alleged particles by measuring the density of wave packets.

Calling it the "measurement problem," Born found that the impact of the measurement would actually produce a "scattering" of the waves in the "wave packet" causing an indeterminacy in the measurement. Producing an unavoidable uncertainty or probability in the measurements in wave mechanics contrary to the strict causality and determinism in classical mechanics, he concluded that this showed it was not closer to traditional physics as Schrödinger claimed. According to Pais: "*It is the first paper to contain the quantum mechanical probability concept.*"[99] In his June paper Born described the scattering by a wave function ψ_{mn}, where the label n symbolizes the initial beam direction, while m denotes some particular direction of observation of the scattered particles. At that point Born introduced quantum mechanical probability: "ψ_{mn} determines the probability for the scattering of the electron . . . into the direction [m]." (p. 286)

In the second paper, published in July, he interpreted Schrödinger's wave function $|\psi|^2$ as the probability for locating the "particle" at the point of greatest density in the wave packet, adding to the measuring probability the probability of quantum states. Although Born had originally believed that Schrödinger's wave mechanics led back to a more traditional interpretation of subatomic physics, his probabilistic interpretations convinced him otherwise. As stated in his autobiography:

> Schrödinger believed . . . that he had accomplished a return to classical thinking; he regarded the electron not as a particle but as a density distribution given by the square of his wave function $|\psi|^2$. He argued that the idea of particles and of quantum jumps be given up altogether; he never faltered in this conviction. . . . I, however, was witnessing the fertility of the particle concept every day in . . . brilliant experiments on atomic and molecular collisions and was convinced that particles could not simply be abolished. A way had to be found for reconciling particles and waves.[100]

Just as Newton's conception of absolute space and time that were based on measurements made by rods and clocks that were unaffected by the relatively slight velocities of the earth had to be revised when Einstein discovered that when approaching the velocity of light measuring rods contract, clocks slow down, and mass increases (to account for the invariant velocity of light), so measurements of the subatomic or quantum world, assuming that they would follow the same Newtonian calculation method, when actually measured, would have to be radically revised.

As usual, physicists were confounded when they encountered the wave-particle duality, the statistical nature of quantum mechanics, and the uncertainty principle due to the interacting measurements at the subatomic level of inquiry that refuted the Newtonian assumption of the universality of the laws of nature at all levels or scales of inquiry. Here again we encounter a further aspect of the third radical revision in our conceptions of reality at different dimensions or levels of inquiry. The bewilderment decreased somewhat when it was discovered that the formalisms of Dirac's theory, Schrödinger's wave mechanics, and Heisenberg's matrix mechanics were equivalent: according to Emilio Segrè, "[f]or all three the essential relation that produces the quantification is $pq - qp = h/2\pi i$ . . . [while] for Heisenberg p and q are matrices; for Schrödinger q is a number and p the differential operator $p = h/2\pi i \, \partial/\partial q$ . . . [and] for Dirac p and q are special numbers obeying a noncommutative algebra. . . ."[101] But the dispute continued with Bohr inviting Schrödinger to his Institute in Copenhagen on October 27, 1926, to discuss their theoretical differences with such intensity that Schrödinger became ill from the tension during the exchange, even though Bohr had the reputation of being a "very considerate and friendly person by nature." Yet no resolution was reached.

After Schrödinger left Copenhagen Bohr carried on his dispute with the same intensity with Heisenberg, who was an associate at his Institute at the time. Trying to resolve their differences with Bohr defending the view that the solution depended on forging the correct *conceptual framework* and Heisenberg insisting, as usual, that the resolution would depend upon devising the correct *mathematical formalism*, they, too, arrived at an impasse. Frustrated and exhausted by these intense discussions, Bohr decided to take a skiing trip to Norway to relax leaving Heisenberg at the Institute to pursue his investigation.

Concentrating on his measurement problem, as a result of his discussion with Bohr, Heisenberg decided to investigate the difficulty involved in measuring the position and momentum of a particle under a gamma ray microscope. The latter is used because its short wavelength provides great accuracy in determining the position, but according to Planck's formula $\varepsilon = h\nu$, a short wavelength also has a high frequency with high energy such that the interaction between the wave and the particle adversely affects the precision of the momentum measurement. To reduce the inaccuracy of the latter a longer wavelength is required, but that produces less certainty in the position measurement.

Rather than trying to remove the discrepancy, by the end of February 1927, Heisenberg had decided to accept it as unavoidable and devise a formula to state what initially came to be known as the famous "uncertainty or indeterminacy relations." An appreciation of the radical change involved is seen if contrasted with what was taken for granted in classical mechanics as stated by Pierre-Simon Laplace in 1886.

> An intellect which at a given instant knew all the forces acting in nature, and the position of all things of which the world consists . . . would embrace in the same formula the motions of the greatest bodies in the universe and those of the slightest atoms; nothing would be uncertain for it, and the future, like the past would be present to its eyes.[102]

It was this assurance that nature is governed by exact laws that would disclose a final knowledge of the universe *at all dimensions* that Heisenberg was rebutting. Having accepted the conjugate indeterminacy, Heisenberg sought a mathematical formula that would describe the resultant uncertainty. Although the conditions necessary for measuring the *conjoined* values of the *conjugate magnitude's* position and momentum, along with energy and time, could not be precisely measured, either of the dimensions alone could be exactly determined, but the more precise the measurement of one the less precise the measurement of the other. As Heisenberg expressed this mathematically: if the *uncertainty* in accuracy of the measurement of each of the interdependent *conjugate* attributes is represented by the delta symbol (Δ), then the product of the *conjoined* magnitudes momentum p and position q cannot be reduced to less than Planck's constant barred, $\Delta p \times \Delta q$ must be equal to or greater than $\hbar$. The second uncertainty states that in the time interval Δt the energy can only be measured with an accuracy equal to or greater than $\hbar$.

Heisenberg published the results in the April 1927 issue of the *Zeitschrift für Physik.* Having received the proofs of the article, Bohr sent a copy to Einstein "adding in an enclosed letter that it 'represents a most significant . . . exceptionally brilliant . . . contribution to the discussion of the general problems of quantum theory.'"[103] What makes it exceptional is not just the calculated mathematical equation, as significant as that is, but that it reversed the age-old assumption that for the mathematics to be correct it must accurately represent the experimental results. Heisenberg affirmed that it is the mathematics that limits or sets the possible experimental outcome! As he states:

> Instead of asking: How can one in the known mathematical scheme express a given experimental situation? the other question was put: Is it true, perhaps, that only such experimental situations can arise in nature as can be expressed in the mathematical formalism? The assumption that this was actually true led to limitations in the use of the concepts that had been the basis of classical physics since Newton.[104]

As an indication of the influence Heisenberg's paper had on Bohr, when a famous article by Einstein, Podolsky, and Rosen (known as the EPR article), titled "Can Quantum Mechanical Descriptions of Physical Reality Be Considered Complete?" was published in the *Physical Review* in 1935,[105] claiming that although Heisenberg's formalism was consistent with all the known quantum data it was "incomplete" because it did not allow precise measurements of the conjugate attributes' position and momentum and energy and time, Bohr had a ready reply.

In the following issue of the *Review*, in an article with exactly the same title, he replied that in quantum mechanics

> we are not dealing with an incomplete description characterized by the arbitrary picking out of different elements of physical reality at the cost of sacrificing other elements, but . . . the *impossibility*, in the field of quantum theory, of accurately controlling the reaction of the object on the measuring instruments. . . . Indeed we have . . . not merely to do with an *ignorance* of the value of certain physical quantities, but with the *impossibility* of defining these quantities in an unambiguous way.[106]

In the draft of a paper in July 10, 1927, Bohr had used the term 'complementarity' for the first time that became a famous designation for the conjugate uncertainty measurements. Then in a collection of articles published later in his life, there is a clear statement of how he believed quantum mechanics has changed our method and understanding of the subatomic quantum domain in contrast to the macroscopic and atomic level of experience, an explanation that had assumed that a precise description and definite explanation of the external world was always possible, even if out of reach at the time:

> Within the scope of classical physics, all characteristic properties of a given object can in principle be ascertained by a single experimental arrangement. . . . In quantum physics, however, evidence about atomic objects obtained by different experimental arrangements exhibits a novel kind of complementary relationship. Indeed, it must be recognized that such evidence which appears contradictory when combined into a single picture is attempted, exhausts all conceivable knowledge about the object. Far from restricting our efforts to put questions to nature in the form of experiments, the notion of *complementarity* simply characterizes the answers we can receive by such inquiry, whenever the interaction between the measuring instruments and the objects forms an integral part of the phenomena.[107]

Just as we found that our sensory system modifies what we observe, so we have learned that what unobservable properties the world discloses experimentally at *a certain dimension* also partially reflects the methods and instruments used in investigating it. This realization that all experience and knowledge is due to an interaction with the world, not just an immediate awareness or disclosure of it as usually appears to be the case, brought about a radical transformation in our conception of reality and how we come to know it—a reversion especially imposed by Heisenberg's discovery of the "uncertainty principle" and Bohr's "Copenhagen interpretation" that has been a crucial feature of the third scientific transformation of our conception of reality.

What is surprising is that Einstein, in his article with Podolsky and Rosen, did not realize that the uncertainties encountered when investigating a deeper domain of particle physics did not permit the same exact measurements as those made on a larger scale was analogous to his special theory of relativity. His theory also claimed that the exact measurements of space, time, and motion made within the lesser velocities of ordinary experience cannot be made when the velocities are so extreme they effect the measuring devices such that the measurements are relative to the velocities of the measurer rather than being absolute as Newton claimed.

For millennia humans believed that the picturesque world as ordinarily experienced was the actual world. Even as late as the nineteenth century Ernst Mach declared that "Atoms cannot be perceived by the senses . . . they are things of thought" implying they did not exist. Yet at the beginning of modern classical science, with the introduction of the telescope and the microscope, scientists began to realize that the existence of the ordinary world, as objective, determinate, and independent as it appears to be, really depends on very complex, unseen underlying conditions. With every discovery of a new dimension of the world the assumption usually has been that this must be the final reality, not just another level of inquiry.

But even if all existence and knowledge is conditional, it is equally erroneous to infer that we do not know anything about the world or that it has any objective properties as concluded in the article by Einstein, Podolsky, and Rosen *if the quantum mechanical worldview were accepted*. But were their view true, how could we account for the corrective and progressive advances in scientific knowledge and its extraordinary technological consequences before and since their time?

Supposing that whatever knowledge of the universe and human existence we acquire depends upon the physical conditions within which they exist, along with the method of investigation used, this does not preclude their being actual within those conditions, otherwise we would have to deny that the ordinary world we live in exists and that the independent subatomic world does not have any of the physical properties it has because their existence is dependent on a more extensive background physical context. Consider water existing as vapor or ice under different conditions.

What we have to realize is that the meaning of 'existence' has changed with the acquisition of greater knowledge, just as Bohr argued that the meaning of 'understanding' has changed. Just because particles are so minute that they *prevent the measuring* of certain conjugate properties does not mean that these properties do not exist in the object conjointly—that the particle does not have a simultaneous position and momentum or energy and time just because the conditions prevent their being measured conjointly. How could it exist without these conjugate properties?

As added evidence of this conception of "contextual realism" that I referred to in a previous book bearing that title and again in this book, I'll continue the review of additional scientific discoveries showing the *limits*, not necessarily the negation, of Newton's corpuscular-mechanistic view of reality at a deeper level of inquiry, along with illustrating that additional physical or quantitative properties of subatomic particles have been discovered despite their existence being dependent upon the type of measurement used to identify them.

The first is the property of spin. It is common knowledge that microscopic particles are not defined by sensory qualities, but by their primary properties of mass, charge, energy, and momentum. At about the time the previously described quantum mechanical discoveries were being made, two physics students in their mid-twenties from the University of Leyden in Holland, Samuel S.

Goudsmit and George E. Uhlenbeck, suggested that on Bohr's model, electrons like planets, in addition to having an orbital motion around the nucleus, also revolve on their axes with an invariant angular momentum called "spin," whose value is ½$h/2\pi$. This value remains constant "even when the electron is outside the atom, and is totally independent of the linear speed or environment of the electron,"[108] indicative of its inherent though conditional nature. Furthermore, because the electron is electrically charged and follows the laws of quantum mechanics, it has two additional properties.

First, having an electrical charge it acts as a tiny magnet whose movement creates a magnetic moment that creates an electromagnetic field. Second, in quantum mechanics an entity with the properties of angular momentum has just two possible spin orientations: "If we perform any measurement whatsoever to determine the angle between the direction of the electron spin and any given direction in space, we find that the angle is *always* either 0^0 or 180^0—in other words, the spin is either parallel or antiparallel to the chosen direction" (pp. 53, 55). Moreover, the spin of certain pairs of subatomic particles are such that measuring the spin of one particle will instantly cause its twin particle to begin spinning in the opposite direction at the same rate, however great the distance between them, a discovery made by John Stewart Bell and published in the *Review of Modern Physics* in 1966. Given these perplexing features of quantum mechanics, it has been claimed that the concept of spin as an *actual* rotary motion of the electron should not be taken literally: "it is more accurate to say that the electron has an intrinsic angular momentum of ½$h/2/\pi$, called spin, *as if* it were rotating about its axis" (p. 57). But how is it possible to use "as if states" in scientific theorizing?

For example, the concept of spin is considered an additional important property of electrons and other particles, having explanatory as well empirical consequences that *have been experi-*

mentally confirmed. The electric charge along with its spin gives the electron its magnetic moment that helps explain the emission and reabsorption of photons. In addition, the two possible spin orientations imply two energy states that help explain "the peculiar pattern of close lines or doublets in the Balmer series of the hydrogen spectrum." Called quantum electrodynamics or QED for short, the theory measured the magnetic moment of the electron "in a unit called the Bohr magneton, denoted μ_e" that has "a very great accuracy . . . found to be 1.001 159 652 μ_e" (p. 58).

As another example of the remarkable influence Bohr's Institute had on the development of quantum mechanics, during the time that Schrödinger and Heisenberg were at the Institute, Paul Dirac also was present doing postdoctoral work from September 1926 to February 1927. Like Schrödinger, he hoped to reconcile quantum mechanics and relativity theory by incorporating Einstein's concept of the field and reformulating Heisenberg's quantum mechanics, along with incorporating the Hamiltonian, the operator corresponding to the total energy of a system. As described by Crease and Mann:

> Using Heisenberg's quantum mechanics, Dirac was able to come up with the Hamiltonian for the atom from quantum mechanics. Dirac was thus able to say that the Hamiltonian for the entire process could be found by adding up the separate Hamiltonians for the atom, the field, and the interaction. . . . The result was the first quantum field theory. Because it linked quantum theory with the dynamics of electromagnetic fields, Dirac called it *quantum electrodynamics.*[109]

He submitted his results in an article to the *Proceedings of the Royal Society* toward the end of January 1927, just three weeks before Heisenberg conveyed his uncertainty principle to him. In succeeding papers he published his well-known relativistic wave equation, which has come to be known as the "Dirac equation" that

Pais says "ranks among the highest achievements of twentieth-century science."[110] In 1928 he devised "a relativistically invariant equation for an electron" whose mathematics "introduced a new internal degree of freedom of the particle. This degree of freedom turns out to have all the properties of the electron spin, starting from its value $h/4\pi$. It also has a magnetic moment of value $eh/4\pi mc$."[111]

As proof of its validity, these properties of spin and magnetic moment were not introduced ad hoc but as properties predicted by the equations. Along with sharing the Nobel Prize with Schrödinger in 1933, in the previous year, as an acknowledgment of his outstanding achievements, Dirac was appointed to Newton's chair of Lucasian Professor of Mathematics at Cambridge University that was occupied by the famous cosmologist Stephen Hawking until his recent retirement.

Yet like medicines that have wonderful curative powers but also unexpected side effects, Dirac's equation produced very puzzling outcomes. For instance, his equation predicted that when the electromagnetic field was quantized and included Heisenberg's uncertainty principle space was no longer empty, but filled with bizarre entities and occurrences, as described by Crease and Mann:

> The spaces around and within atoms, previously thought to be empty, were now supposed to be filled with a boiling soup of ghostly particles. From the perspective of the quantum field theory, the vacuum contains random eddies in space-time: tidal whirlpools that occasionally hurl up bits of matter, only to suck them down again. Like the strange virtual images produced by lenses, these particles are present, but out of sight; they have been named *virtual particles.* Far from being an anomaly, virtual particles are a central feature of quantum field theory, as Dirac himself was soon to demonstrate.[112]

But as peculiar as these predictions were, there was another just as weird. When the Hamiltonian (energy) of a single electron was predicted from his equation it showed two possible values, one negative and one positive. The existence of positive energy was of course well-known, but no one had ever encountered negative energy. Also, when using his improved Wilson cloud chamber to detect cosmic waves Carl D. Anderson observed tracks of what appeared to be light particles. On further investigation to determine whether they were negatively or positively charged, he found they were positive. Thus he accidentally discovered a new particle with a mass comparable to an electron but with an opposite positive charge that he named 'positron,' after the electron. Given their opposite charges, when they interact they annihilate thereby producing two photons. Consequently, he had discovered a new kind of matter called "antimatter." According to Crease and Mann:

> From an embarrassment the negative energy states were transformed into a triumph for quantum electrodynamics, the first time in history that the existence of a new state of matter had been predicted on purely theoretical grounds. Dirac won the Nobel Prize in 1933; Anderson went to Sweden three years later. (p. 90)

In the following decades Dirac's quantum electrodynamics (QED), consisting of six quarks, six leptons, and five bosons, was developed into what became known as "the standard model" independently identified by three physicists: the Japanese born Sin-itiro Tomonaga and two Americans, Richard Feynman and Julian Schwinger, all three receiving the Nobel Prize for their achievement in 1965. (I believe it was on July 4 or the 5, 2012, that the scientists at the European Organization for Nuclear Research (CERN) in Geneva announced the discovery of the Higgs boson, or "God particle" as it is now called.) Applying quantum mechanics to electromagnetic fields and to electrons (with the wave function of the electron also considered a field), they treated the fields not as a continuum but composed of discrete quanta.

According to Chris Quigg at the Fermi National Accelerator Laboratory (Fermilab):

> QED is the most successful of physical theories. Using calculations . . . developed in the 1940s by Richard P. Feynman and others, it has achieved predictions of enormous accuracy, such as the infinitesimal effect of the photons radiated and absorbed by the electron on the magnetic moment generated by the electron's innate spin. Moreover, QED's descriptions of the electromagnetic interaction have been verified over an extraordinary range of distances, varying from less than 10^{-18} meter to more than 10^{8} meters.[113]

Continuing research on the magnetic moment, in 1933, after enhancing the original Stern-Gerlach experiment, Stern found that the magnetic moment of the proton was three orders smaller than that of the electron and that of the neutron, and despite being neutral has a "negative magnetic electric charge" similar to the proton. As typical of scientific inquiry, these "facts hinted that the neutron (and also the proton) have an internal structure which includes positive and negative charges, because magnetism always involves the motion of charges."[114]

Additional contributions were made by Wolfgang Pauli that include the introduction of the "exclusion principle," the hypothesis of the "neutrino," and the importance of spin in determining which particles and in what number can occupy an atomic orbit. An Austrian physicist with an unusually critical and acerbic manner who occasionally signed his communications with "The Wrath of God," Pauli nonetheless was a gifted scientist who gained the esteem of his colleagues. His major contribution, "the exclusion principle" introduced in February 1925, added to Bohr's earlier explanation of the limitation of the kinds and numbers of electrons that could occupy the successive stationary orbits in his solar model of the atom. For this he was awarded the Nobel Prize in physics in 1945.

> In due course it was found and proven theoretically that the Pauli exclusion principle is valid for any particle whose spin is not integral, i.e., [whose spin is] [1/2, 3/2, 5/2, etc. The laws of behavior of these particles are embodied in "Fermi-Dirac statistics." . . . These are the statistics characterizing distinguishable objects. The particles themselves are called *fermions*. . . . Protons and neutrons are also fermions (spin ½) and thus in a nucleus they populate different energy levels just as the electrons in the atom do. (p. 60; brackets added)

In contrast, particles with integral spin or whole numbers are called "bosons" (after the Indian physicist S. N. Bose who, along with Einstein, identified them) and are not affected by Pauli's exclusion principle. Not being distinguishable by the four quantum numbers, they are governed by another kind of statistics named the "Bose-Einstein statistics" and since they are capable of having the same quantum numbers an unlimited amount can be located in a particular region of space. "It can be shown that the difference between fermions and bosons is related to the connection between the spin and the symmetry of the wave function of the particles" (p. 60).

Along with the discovery of the spin vector and quantum numbers, physicists were attempting to determine the sizes of the various particles that the wave-particle duality with its contrasting properties, along with the uncertainty principle with its obscurity, made particularly difficult. In the thirties physicists tried to measure the diameter of the electron and "arrived at the formula $r = e^2/mc^2$ where e and m are the charge and mass of the electron and c the speed of light. This gave a value of 3×10^{-13} centimeters for the radius" (p. 61). But calculations made within the system QED, where the electron is considered a mere point, showed a more miniscule value of 10^{-16} centimeters while experiments on protons and neutrons "show that their mass and charge concentrated in a region with a diameter of about 1.2×10^{-13} centimeters (p. 61).

I shall now endeavor to present as lucidly and comprehensively as possible the subsequent discoveries of the major subatomic particles and forces that culminates the third revolutionary scientific development that not only extended or replaced Newtonian classical science, but also led to the creation of the contemporary conception, called "Quantum Chromodynamics," of the inner composition of the subatomic particles previously mentioned. This proved exceedingly difficult because of the illusiveness and vagueness of the experimental evidence and the much greater reliance on mathematics. As a result the two major contributors to the new "Standard Theory," Richard Feynman and Murray Gell-Mann, were at times in agreement and at times quite opposed, though it was Gell-Mann's interpretation that usually prevailed. His contributions to particle physics were extraordinary: the property of strangeness,V-A, the Eightfold Way, quarks (although for a long time he wavered as to whether they were real or just artifacts to preserve the mathematical symmetry), quantum chromodynamics (QCD), along with many others discoveries too numerous to cite.

The succeeding decades of the twentieth century following the Second World War, owing to the creation of the atomic bomb and the increasingly powerful atomic accelerators, such as the Large Hadron Collider at CERN, near Geneva; the Fermi National Accelerator Laboratory (FNAL) in Batavia, Illinois; the Brookhaven National Laboratory; and the Stanford Linear Accelerator (SLAC) produced the detection of a deeper domain of subatomic particles and forces by accelerating and colliding a deeper level of particles. Created out of the mass-energy equivalence stated in Einstein's formula $E = mc^2$, they were predicted and/or discovered by such outstanding scientists as Eugene Wigner, Richard Feynman, Murray Gell-Mann, Julian Schwinger, Steven Weinberg, Sheldon Glashow, Hideki Yukawa, Samuel Ting, Burton Richter, Harald Fritzsch, and many others.

Such a mélange of new particles have been discovered along

with the previously mentioned basic particles, such as the proton and neutron, that they have been compared to a zoo and required the creation of a new periodic table composed of hadrons with the hadrons further divided into baryons and mesons. Two new forces were added to gravity and electromagnetism, a strong force binding the nucleons and a weak force explaining the radioactivity within the nucleus that only acts within short distances. Furthermore, previous forces acting at a distance were superseded by an exchange of "virtual particles," photons in electromagnetism, gluons in strong interactions, and the vector bosons W^-, W^+, and Z^o in weak interactions. While hadrons react to both strong and weak forces, leptons only respond to the weak force consisting of the exchange of three bosons and photons.[115]

As usual, at first it was thought that the hadrons and leptons, along with the photons and hypothesized gravitons completed the list of basic particles, but in 1964 Murray Gell-Mann and George Zweig introduced further particles accounting for the structure of the hadrons that Gell-Mann whimsically named "quarks" (taken from a passage in James Joyce's *Finnegans Wake*, "Three quarks for Muster Mark!"), that caught on! They do not have integral charges but fractional charges of plus two-thirds or minus one-third. And like the ancient "minima" of Epicurus, they never exist separately but are conjoined as pairs or triplets to form hadrons, a quark and an anti-quark comprising mesons and three quarks forming a baryon. Although originally they were just conjectures to explain the interactions of the hadrons, their depicted combinations into hadrons have been verified.

Continuing the discoveries endowed with fanciful names, two new charges were postulated, one called "strangeness" by Gell-Mann and another referred to as "charm" by Glashow, enabling physicists to account for hadrons and their interactions based on combinations of quarks classified as "flavors": up (u), down (d), charm (c), strange (s), bottom/beauty (b), and top/truth (t) (pp.

291–95). Thus matter was classified into two groups, one consisting of six leptons and another of six quarks. Then a new quantum theory analogous to quantum electrodynamics (QED), based on the Yang-Mills gauge theory and group theory, was conceived and named "quantum chromodynamics" (QCD for short) by Gell-Mann (p. 291). It was so named because it consisted of a strong force or charge called "color" that binds the quarks within the hadrons.

Further unifications occurred when Glashow combined the weak and electromagnetic forces into an "electroweak theory" and Stephen Weinberg introduced the concept of "symmetry breaking" by the (postulated) Higgs boson introduced to explain how Glashow's W and Z particles acquired mass, the existence of which has been recently confirmed to the great satisfaction of nuclear physicists. The unification continued when in 1969 the Dutch physicist Gerardus 't Hooft discovered that Cartan's group theory could be applied to gauge theories that allowed Weinberg in 1973, utilizing the color charge, to create "a gauge field theory of strong forces," the basis of QCD (pp. 278–79).

But the weak forces were still unaccounted for. Thus Glashow and James Bjorken wrote a paper in which they suggested that a new quark called "charm" could link QCD with the electroweak theory, despite the difficulty of its confirmation because quarks do not exist independently, but only in self-enclosed couplets or triplets within the hadrons. Yet their existence was eventually confirmed. And so the discoveries continued with the detection by Samuel Ting at Brookhaven and Burton Richter at SLAC of a new particle called "J" by Ting and "psi" by Richter, now known as the "J/Y particle." They were awarded the Nobel Prize in 1976 for their discovery, and Glashow, Weinberg, and Salem shared the prize in 1979 for contributing to the theory of electroweak and electromagnetic interactions between elementary particles. And as a kind of culmination, Carlo Rubbia and Simon Van der Meer were awarded the Nobel Prize in 1984 for their confirmation at CERN of the existence of the

W^-, W^+, and Z^0 particles. Feynman, Schwinger, and Tomonaga had been awarded the Nobel Prize in 1963 while six years later in 1969 Gell-Mann was the single recipient of the prize for his numerous outstanding contributions to physics.

Despite these significant advances, the attempt to unify the strong and weak forces had just begun. Glashow and Howard Georgi advanced the effort toward unification in a series of papers published in 1973–1974, one of which carried the impressive title "Unity of All Elementary-Particle Forces." If successful, their Grand Unified Theory (GUT as it came to be called) would combine quarks and leptons into one family, owing to their decaying into one another, while a "superweak force" was introduced to unify strong and electroweak interactions. Later in the same year Georgi, Weinberg, and Helen Quinn wrote a paper declaring that at very high temperatures or energies, such as existed at the time of the Big Bang, all the forces were unified. While not yet confirmed, at least the initial framework of a Grand Unified Theory was constructed.

In an *Atlantic Monthly* article published in 1984 with the bold title "How the Universe Works," Crease and Mann summarized these achievements as follows:

> The result is a ladder of theories. Firmly on the bottom is SU(3) [a Family of eight baryons Gell-Mann again fancifully designated the "eightfold way" after Buddha] × SU(2) × SU (1) [the SU stands for special unity group based on Cartan's group theory], whose predictions have been confirmed ("to the point of boredom," Georgi says). . . . The W and Z particles were discovered at CERN . . . but the theory was so well established by then the event was . . . anticlimactic.
>
> The GUTS proposed by Georgi and Glashow and other physicists, that fully unite the strong, weak, and electromagnetic forces, are the next rung on the ladder. Although as yet unconfirmed . . . these theories are considered compelling by

> most physicists. Finally, at the top of the ladder, in the theoretical stratosphere, are supersymmetry and its cousins, which are organized according to a principle somewhat different from SU(5), though, like that model, they put apparently different particles together in groups. Supersymmetry groups are large enough to include gravity, but are so speculative that many experimenters doubt they can ever be tested.[116] (brackets added)

Although published many years ago and therefore somewhat dated, the quotation presents an excellent summary of developments up to that time showing how radically different the scientific framework had become since the time of Newton. Not only is the world no longer completely deterministic and limited to the atomic domain, the recent advances in physics have become so dependent on the mathematical formalism that it is impossible to render it intelligible in more familiar, pictorial, or visualizable concepts. Perhaps an exaggeration, but in 1992 Weinberg skeptically claimed like Einstein in the EPR article that

> quantum mechanics by itself is not a complete physical theory. It tells us nothing about the particles and forces that may exist. Pick up any textbook on quantum mechanics; you find as illustrative examples a weird variety of hypothetical particles and forces, most of which resemble nothing that exists in the real world, but all of which are perfectly consistent with the principles of quantum mechanics. . . . Most of these theories can be logically ruled out because they would entail nonsense like infinite energies or infinite reaction rates.[117]

Yet perhaps influenced by the constant references to the experimental confirmation of new discoveries and explanations, I found that most of the physicists at the time were reminiscent of John Trowbridge who had discouraged students at the beginning of the twentieth century from pursuing graduate work in physics believing that nearly everything of importance had been explained. In the latter decades of the century most physicists thought that the final answers would be found in the near future. As Harald Fritzsch, a coworker of Gell-Mann's, wrote in 1943 in the original German then translated into English in 1983 with no retraction:

> Today we can state unequivocally that the physics of the atom is understood. A few details need to be cleared up, but that is all. . . . Important and fascinating discoveries have been made in high energy physics, especially since 1969, and today it appears that physicists are about to take the important leap toward a complete understanding of matter.[118]

Also in his inaugural lecture in April 1980, titled "Is the End in Sight for Theoretical Physics?" when he assumed the Lucasian Chair in physics at Trinity College at Cambridge University, Stephen Hawking, the world's most famous cosmologist, had the following read (because he has amyotrophic lateral sclerosis):

> *In this lecture I want to discuss the possibility that the goal of theoretical physics might be achieved in the not-to-distant future, say, by the end of the century. By this I mean that we might have a complete, consistent, and unified theory of the physical interactions which would describe all possible observations. Of course one has to be very cautious about making such predictions. . . . Nevertheless, we have made a lot of progress in recent years and . . . there are some grounds for cautious optimism that we may see a complete theory within the lifetime of those present here.*[119]

In fairness, owing to developments in physics since then, he has lately rescinded his statement. In his recent book previously referred to, *The Grand Design*, written with Leonard Mlodinow, Hawking describes the cosmological theory of "multiuniverses," which would contain a vast number of alternate universes whose laws of nature would be very different from ours, thus making a final theory extremely difficult or unlikely.[120] Still, as Johnson states in *Strange Beauty*:

> Whatever one's philosophical inclinations, it was hard not to be in awe of the Standard Model. Discovery or invention, it was a work of art. Whether it was the art of nature or the art of human kind could never be known for sure. But it was pleasing to think that humans, on their tiny planet, with their blinkered senses and animal brains, could weave observation and imagination into such a powerful theory. (p. 296)

Before concluding this discussion of the accomplishments of the second revolution I should mention the accurate discovery of the age of the universe since the Big Bang. Hubble had detected the recession of the galaxies from the earth thereby indicating the expansion of the universe, and devised an equation to calculate the rate of the recession as "$Ho = v/d$ (where Ho is the constant, v is the recessional velocity of a flying galaxy, and d its distance away from us)."[121] Based on his formula he calculated the age of the universe to be "about two billion years old." It wasn't until "February 2003 that NASA and the Goddard Space Flight Center, using a new type of satellite called the Wilkinson Microwave Anisotropy Probe, announced the age of the universe as 13.7 billion years . . ." (pp. 169–70). Just recently the European Space Agency, using their Planck telescope, determined the age of the universe to be 13.8 billion years. It was in 1953, after many other unsuccessful efforts, that Clair Patterson by measuring the amount of radioactive decay of uranium to lead in ancient earth rock crystals, determined the age of the earth to be about "4,550 million years (plus or minus 70 million year)—a figure that stands unchanged 50 years later . . ." (p. 157).

Chapter VIII

THE IMPENDING FOURTH TRANSITION ALONG WITH THE FUTURE PROSPECTS OF SCIENCE

Having described the third revolution, the challenge now is to access what physics portends for the future. In the last decades of the twentieth century the main problems confronting physicists were the unification of the electroweak and electromagnetic forces, forging a more definitive model of the atom, and confronting the revisions in physics owing to the discovery of the strange quantum of energy. Along with other outstanding contributors, two contrasting worldviews were offered as solutions by the two most influential physicists of the century, Einstein and Bohr.

Despite being the originator that light and matter consist of swarms of quanta that raised all the perplexing questions as to if and how they compose physical reality, Einstein was adamant in his defense of an eventual interpretation based on the discoveries of the *underlying objective properties* of an independently existing physical world. As he wrote: "The belief in an external world independent of the perceiving subject is the basis of all natural science."[122]

In contrast, Bohr defended the view that given the impossibility of observing quanta due to their infinitesimal size it was necessary to rely on measurements based on the *interactions* of the measuring instruments with the quanta, despite realizing that as a result the measured data were not inherent properties of independently existing particles as normally believed, but "complementary" because they could not exist separately. It also meant that the measurements were uncertain and probable and that the quantum itself did not exist until measured! As Bohr stated: "There is no quantum world. There is only an abstract quantum mechanical description. It is wrong to think that the task of physics is to find out how nature is. Physics concerns what we can say about nature."[123]

This limitation to the measurable data led to what has been called the "Copenhagen Interpretation" of Quantum Mechanics named after Bohr's Institute whose extremely brilliant visiting students and scholars, usually with his guidance, played the leading role in the development of quantum mechanics. The interpretation was that since all the quantum evidence was based on and limited to the mathematical measurements the resultant *experiential phenomena* composing the theory consisted solely of the system of measurements without representing or being dependent on any independent underlying physical reality as it's cause! This in turn led to fantastic interpretations by psychedelics, Eastern mystics, and Christians who claimed that it was the direct word of God.

Despite its having no empirical content, it nonetheless helped create many of the outstanding technological innovations of the twentieth and twenty-first centuries, such as radios, television, computers, mobile phones, iPod®s, and quantum cryptography, leading most physicists to accept it despite its obscurity and tenuousness. Although Bohr's Copenhagen Interpretation claimed that there was no possibility of a credible theoretical interpretation, several new theories were introduced such as a "Cosmic Multiverse," an "Inflationary Universe," "Parallel Worlds," and "String Theory." The first three present different explanations of the inflationary universe or that in addition to our known universe there are multiple universes each with its own unique laws. Since it is not practical to discuss and evaluate all four theories, I will restrict my discussion to String Theory, which is perhaps the best known and that has been the subject of intensive research for about four decades and yet still has not produced any supporting experimental evidence, even though nearly all the predicted—though probable—experimental evidence for quantum mechanics, as bewildering as it is, has been confirmed, which is the reason so many physicists accept it as a final state.

Yet the confirmation of other new theories continue to be announced sustaining a certain confidence in the continuing progress of physics. The *Washington Post* (July 5, 2012) published an article subtitled, "Subatomic Particle Is Breakthrough in Understanding [the] Basics of the Universe" declaring that "scientists in Europe [announced] that they'd found the Higgs boson or something remarkably Higgs-like, [that] was a stunning triumph of both theory and experiment" (brackets added).[124] The particle was so crucial that it was called the "God particle" by Leon Lederman, its discovery hailed in the article as "a tremendous breakthrough with enormous explanatory significance."

Yet the particle itself has never been *directly* detected, the evidence for its existence inferred from the confirmation of what has been mathematically predicted characteristic of quantum mechanical inquiries. As the authors state:

> Regardless of whether it's the Higgs or a Higgs imposter, it's a very real particle, and newly known to science, and apparently fundamental to the texture of the universe. . . . It was also the most important, because it is thought to give rise to the "Higgs field," a sort of force field that permeates everything. (pp. A1, A2)

Whether this is true or not is one of the crucial questions facing physicists today. Still, many have retained their confidence in their ability to solve such problems, as implied in the statement by Michael S. Turner, a professor of physics at the University of Chicago, at the end of the Higgs article: "Okay, the particle physicists got their Number 1 wish—the Higgs. Now we cosmologists want ours—the dark-matter particles."

Although it is undeniable that extraordinary progress has been made since the seventeenth century in understanding those domains of the universe that we have attained access to, is it realistic to assume, given the duration, extent, and complexity of the universe, especially in contrast to the finitude of human existence, that we are capable of attaining a final or complete understanding of the cosmos as most physicists aspired to and many still do? A number of diverse views regarding the future prospects of physics and cosmology have been published, such as *The Shaky Game: Einstein Realism and the Quantum Theory*, by Arthur Fine (1986); *The End of Physics: The Myth of a Unified Theory*, by David Lindley (1993); *Dreams of a Final Theory*, by Steven Weinberg (1993); *The End of Science: Facing the Limits of Knowledge in the Twilight of the Scientific Age*, by John Horgan (1996); *The Closing of the Western Mind: The Rise of Faith and the Fall of Reason*, by Charles Freeman (2003); *The Beginning of Infinity: Explanations that Transform the World*, by David Deutsch (2011); and, *Physics of the Future: How Science Will Shape Human Destiny and Our Daily Lives by the Year 2100*, by Michio Kaku (2012).

However, as discussed in the previous chapter, that the existence of the observed world, along with the subatomic realm and especially the even more basic quantum domain that had been accepted by many scientists as the "standard model," are now believed to depend solely upon the instruments of detection and the type of mathematics used, this does not necessarily mean that they have no "objective" nor "real status" independently of the method of inquiry. However bewildering quantum mechanics seems to be, it represents the only credible view of physical reality at that level of inquiry, but not the final truth.

How could the Periodic Table have been organized; contagious epidemics eliminated by inoculations; Hubble's first telescope built at Mount Wilson's Observatory that disclosed the existence of additional galaxies and the accelerating inflationary universe; Hubble's later space telescope followed by the James Webb Space Telescope that revealed trillions of stars and exoplanets existing in about 125 billion galaxies, enabling astronomers eventually to examine the exoplanets more closely to determine whether intelligent beings exist on those other planets, if nothing we have learned about the universe were true? In addition, the new technology includes nuclear reactors; computers and artificial intelligence; an enormous cyclotron constructed to simulate the early physical conditions of the big bang; and the decoding of the human genome. Any prognoses of the future of science must acknowledge our past and impending achievements, even if attaining a conclusive, all encompassing final theory may be beyond our reach.

Since quantum mechanics especially has risen in significance in scientific inquiry with an increasing reliance on a mathematics with inherent uncertainties and probabilities, this has reduced the traditional definiteness, firmness, and pictographic nature of theories and explanations making them more conditional, contingent, and obscure. It is this that has led to the disbelief expressed by David Lindley and also by Gregory Chaitin, as quoted by John Horgan in *The End of Science*, who denounced real numbers as "nonsense" declaring that "Physicists know that every equation is a lie . . . ."[125] But surely this is not true!

How could terrestrial motions be predicted with accuracy if Newton's formula F = ma were a lie; the atomic and hydrogen bombs have been constructed if Lisa Meitner's explanation of uranium fission or Einstein's equation $E = mc^2$ were false; or the successful landing of the rover Curiosity on Mars if all the precise calculations were erroneous? Like Kant's assertion that to know the independent "noumenal world" we would have to know "things as they are in themselves," which he declared impossible, Lindley also denied that we have knowledge of "an objective, real world," in opposition to Einstein's assertion, quoted previously, that "[to] believe in an external world independent of the perceiving subject is the basis of all natural science."

It seems to me that there is considerable evidence that Einstein's assertion has been vindicated, as in our theory of electromagnetism, light waves and photons, charged particles such as the electron, proton, neutron, and positron, strong and weak nuclear forces, gluons, gravitons, quarks, the structure of biological cells, evolutionary theory, the discovery of the helical design of the genome, and the Big Bang theory of the universe. Among the things we still do not understand is what caused the massive concentration of energy known as the "big bang" or what preceded it. Was it just an offshoot of multiuniverses perhaps having inherently different laws? If the universe, as some claim, is composed of 23% dark matter (including dark holes), about which little is known, and 72% of equally mysterious dark energy (but considered the expansive force that counteracts gravity causing the earth to recede at an accelerated rate) totaling 95%, that leaves only 5% of the universe of which we have any understanding on which to base a definitive conclusion.

Yet as is usual in scientific inquiry, there now is an attempt to discover the nature of this dark energy believed to surround every galaxy, surmising that it is composed of subatomic particles called WIMPs standing for "weakly interacting massive particles." As described in an article in the *Washington Post*:

> The idea is to hang densely packed strands of DNA — quadrillions per layer — from thin sheets of gold foil. When a dark matter particle smacks into a gold atom, it would knock the nucleus through the DNA, shearing strands as it goes. Researchers could figure out *the path the particle traveled* by seeing where the strands were cut.[126]

Katherine Freese, one of the theoretical physicists involved in the investigation, said that "if the detector works, finding evidence of just 30 WIMPs will be enough to prove that these elusive particles do, in fact, exist. . . . An 80-year cosmic mystery solved" (E-5). Yet some cosmologists hoping to achieve a greater unification, along with a simpler and more elegant theory perhaps compressible into a single equation, as Einstein aspired to, have created string or superstring theory as the solution. Michio Kaku's dissertation topic being on problems in Gabriele Veneziano and Mahiko Susuki's string theory introduced in 1968, along with his collaboration with Keiji Kikkawa at Osaka University, enabled him to "successfully extract" from "the field theory of strings . . . an equation barely an inch and a half long" that "summarized all the information contained within string theory"[127] shows his own involvement in working out the theory.

Attempting to explain all the diverse particles and forces of recent physics in terms of a deeper level of reality, they have returned, ironically, to the ancient Greek philosophy of Pythagoras who conceived of the universe as a musical harmony. As Kaku states:

> If string theory is correct . . . the link between music and science was forged as early as the fifth century B.C., when the Greek Pythagoreans discovered the laws of harmony and reduced them to mathematics. They found that the tone of a plucked lyre string corresponded to its length. If one doubled the length of a lyre string, then the note went down by a full octave. If the length of a string was reduced by two-thirds, then the tone changed by a fifth. Hence, the laws of music and harmony could be reduced to precise relations between numbers. . . . Originally, they were so pleased with this result that they dared to apply these laws of harmony to the entire universe. Their effort failed because of the enormous complexity of matter. However, in some sense, with string theory, physicists are going back to the Pythagorean dream. (p. 198)

As there is no point in my trying to paraphrase Kaku's excellent description of string theory, I will quote his own account.

> According to string theory, if you had a supermicroscope and could peer into the heart of an electron, you would see not a point particle but a vibrating string. (The string is extremely tiny, at the Planck length of 10^{-33} cm, a billion billion times smaller than a proton, so all subatomic particles appear pointlike.) If we were to pluck this string, the vibration would change; the electron might turn into a neutrino. Pluck it again and it might turn into a quark. In fact, if you plucked it hard enough, it could turn into any of the known subatomic particles. In this way, string theory can effortlessly explain why there are so many subatomic particles. They are nothing but different "notes" that one can play on a superstring. . . . The "harmonies" of the strings are the laws of physics. (pp. 196–97)

Moreover, he believes the theory also can explain their interactions and most of the problems of theoretical physics.

> Strings can interact by splitting and rejoining, thus creating the interactions we see among electrons and protons in atoms. In this way, through string theory, we can reproduce all the laws of atomic and nuclear physics. The "melodies" that can be written on strings correspond to the laws of chemistry. The universe can now be viewed as a vast symphony of strings. (p. 197)

He goes on to show how field string theory can explain Einstein's special and general theories of relativity and even provide a possible explanation of the "riddle of dark matter" and "black holes." But while he asserts that string theory can "effortlessly explain" the creation and interaction of the basic physical particles and many other puzzles confronting physics, one can understand why none of it has ever been confirmed. To me it reads like scripture in which declarations are presented with a kind of doctrinal authority, but based on mathematics rather than revelation.

It not only seems unlikely that minuscule vibrating strings can be the source of the universe, the theory requires a multi-dimensional hyperspace for their existence.

> Only in ten-or eleven-dimensional hyperspace do we have "enough room" to unify all the forces of nature in a single elegant theory. Such a fabulous theory would be able to answer the eternal questions: What happened before the beginning? Can time be reversed? Can dimensional gateways take us across the universe? (Although its critics correctly point out that testing this theory is beyond our present experimental ability, there are a number of experiments currently being planned that may change this situation. . . . (p. 185; italics added)

He goes on to discuss refinements in the theory, such as "supersymmetry," "M-theory," "heterotic string theory," and the "Brane World," along with the possible experiments being planned to confirm it, though no supporting empirical evidence that I know of has been announced since 2005 when Kaku's book was published.

Yet despite his somewhat optimistic assessment of the theory, he seems to have the same reservations about it that I have expressed, as the following quotation indicates. Recalling Pauli's version of the unified field theory that he developed with Werner Heisenberg described by Niels Bohr as "crazy," but not "crazy enough," Kaku states:

> One theory that clearly is "crazy enough" to be the unified field theory is string theory, or M-theory. String theory has perhaps the most bizarre history in the annals of physics. It was discovered quite by accident, applied to the wrong problem, relegated to obscurity, and suddenly resurrected as a theory of everything. And in the final analysis, because it is impossible to make small adjustments without destroying the theory, it will either be a "theory of everything" or a "theory of nothing." (pp. 187–88)

Given how much is still unknown or conjectured about the universe, how likely is it that we are close to a "final theory of everything" that would resemble string theory, or that it is even attainable? I have recently read Marcelo Gleiser's work titled *The Island of Knowledge*, published in 2014 after I had written my book, thus I did not have the benefit of reading his exceedingly informed and, in my opinion, correct interpretation of the current controversy in physics, as to whether quantum mechanics represents a realistic and final account of physical reality. His conclusion is that "Unless you are intellectually numb, you can't escape the awe-inspiring feeling that the essence of reality is unknowable" (p. 193), although there is no sounder method of inquiry now than science. While one can concede that quantum mechanics is in a sense *correct* in that it mainly agrees with the current experimental evidence, this does not mean that it is true and is thus a *final theory* of reality. I strongly recommend reading Gleiser's book if one is interested in the prospects of quantum mechanics.

In addition to the question of whether it is presumptuous or realistic to suppose that finite creatures living in this infinitesimal speck and moment of the universe will ever arrive at a final theory, there is the additional problem of whether we can afford the tremendous costs of further research. The discovery of the Higgs or Higgs-like boson cost 10 billion dollars involving 6,000 researchers and the creation of a 17-mile circular tunnel under the border of France and Switzerland with thousands of magnets. The international fusion mega-project now in construction in southern France is estimated to cost 23 billion dollars and whose completion is projected to take a decade. Even continuing research on whether WIMPS exist will ultimaely depend upon the costs, as well as the experimental and theoretical ingenuity.

As examples of how difficult it has become to finance such projects, in 1993 the US Congress discontinued the financing of the superconducting, supercollider after already spending 2 billion dollars digging a tunnel 15 miles long in Texas in which to house it on the grounds that the cost of completing the project was too great. Given the current economy, President Obama has requested a budge cut for our fusion research by 16% to 248 million dollars, a foreboding sign of the future.

I am not suggesting that funding scientific research has not paid off; just consider all the technological, economic, social, medical, and intellectual benefits derived from scientific inquiry to acknowledge the opposite. Everything we know about the universe and human existence and all the economic, educational, social, and medical improvements and advances in our standard of living we owe entirely to the genius and dedication of scientists. But one can't help wondering whether the cost of delving further into the universe will overreach at some point our financial assets and/or capacities. Thus, though I admire most of what he says in his very stimulating book, I question Alex Rosenberg's confident assertion that "Physics is *causally closed* and *causally complete.* The only causes in the universe are physical. . . . In fact, we can go further and confidently assert that the physical facts *fix* all the facts."[128] If true, this would confirm Einstein's worldview. I wish I could be so confident.

Having expressed my reservations that attaining a final theory of the universe is within reach or even possible, I will conclude this study of the major transitions in our conceptions of reality and way of life by citing the amazing scientific and technological advances that are predicted to take place by the end of this century or the next based on the knowledge already attained or anticipated. This, fortunately, has also been comprehensively described by Michio Kaku in his prophetic book previously cited, *Physics of the Future: How Science Will Shape Human Destiny and Our Daily Lives by the Year 2100.* As described on the back cover of the book:

> Renowned theoretical physicist Michio Kaku details the developments in computer technology, artificial intelligence, medicine, and space travel that are poised to happen over the next hundred

> years . . . interview[ing] three hundred of the world's top scientists—already working in their labs on astonishing prototypes. He also takes into account the rigorous scientific principles that regulate how quickly, how safely, and how far technologies can advance . . . forecast[ing] a century of earthshaking advances in [science and] technology that could make even the last centuries' leaps and bounds seem insignificant.[129] (brackets added)

An unexpected and exceptional added attraction of the book is his occasional indication of how the extraordinary modern scientific and technological achievements have often replicated the divine exploits attributed to the gods in ancient mythologies and current religions, such as creating miraculous cures and performing marvellous feats like conferring on humans supernatural powers and eternal life.

The challenge is to present as briefly, clearly, and objectively as possible the range of the incredible developments, greater than the Industrial Revolution, that are predicted to radically change the conditions and nature of human existence in this century or the next, and try to discriminate between the fanciful and the possible, along with the beneficial and harmful outcomes. According to Kaku, one of the basic factors driving this process is the rapidity in the development of computers and how this has altered our lives owing to what is called Moore's law.

> The driving source behind . . . [these] prophetic dreams is something called Moore's law, a rule of thumb that has driven the computer industry for fifty or more years, setting the pace for modern civilization like clockwork. Moore's law simply says that computer power doubles about every eighteen months. First stated in 1965 by Gordon Moore . . . this simple law has helped to revolutionize the world economy, generated fabulous new wealth, and irreversibly altered our way life. (p. 22; brackets added)

The technological developments that were most instrumental in creating the computer revolution apparently were the following: (1) by relying on electrical circuits computers could perform close to the speed of light that permits nearly instantaneous transmissions and communication with the rest of the world; (2) these electrical conductions are further enhanced by the development of miniaturized transistors or switches; and (3) the creation of the computer chip or silicon wafer the size of one's fingernail that can be etched with millions of tiny transistors to form integrated units making it possible to carry out instantaneously enormously intricate calculations that, otherwise, would have taken decades, years, or even centuries.

Turning now to Kaku's account of the various conceptions and predictions of the future developments that will be brought about by the computer revolution, the one I find the most startling and threatening is based on computerized artificial intelligence and the creation of robots that in the most extreme case could, it is *predicted,* replace or convert human beings into computerized robots, as indicated in the initial section chapter 2 of his book *The End of Humanity?* (p. 75).

As of now the most advanced robot is ASIMO created by the Japanese "that can walk, run, climb stairs, dance, and even serve coffee" and "is so lifelike that when it talked, I half expected the robot to take off its helmet and reveal the boy who was cleverly hidden inside" (p. 77). In addition, there "are also robot security guards patrolling buildings at night, robot guides, and robot factory workers. In 2006, it was estimated that there were 950,000 industrial robots and 3,540,000 service robots working in homes and buildings" (pp. 87–88). But while these are remarkable achievements, they are not indications that the robot has attained an ounce of control over or initiates any of its behavior. Everything ASIMO does has been preprogrammed so that its actions are entirely beyond its control. It of course has no conscious awareness of its surroundings or any feelings since every action it performs is computerized. In some cases it is controlled by a person who directs the actions from the images on a computer thousands of miles away, similar to controlling a drone.

More remarkable was the event in 1997 when "IBM's Deep Blue accomplished a historic breakthrough by decisively beating world chess champion Gary Kasparov. Deep Blue was an engineering marvel, computing 11 billion operations per second" (p. 80). Nonetheless, Deep Blue cannot take credit for the achievement that has to be attributed to the intelligence of the gifted programmers who devised all the correct moves to beat Kasparov.

This fact was not lost on the artificial intelligence (AI) researchers who then began attempting to "simulate" conscious awareness by installing object recognition, expressing inner emotional states and feelings by facial expressions, and initiating intelligent actions. Thus, instead of the top-down approach of treating robots like digital computers with all the rules of intelligence preprogrammed from the very beginning, they began imitating the brain's bottom-up approach. They tried to create an artificial neural network with the capacity of learning from experience that would require conscious awareness of the environment, along with emotions and affective feelings that are the source of value judgments, such as whether things are beneficial or harmful.

In addition to attempting to replicate the learning process of human beings, they would have had to install such mental capacities as memory, conceptualizing, imagining, speaking, learning languages, and reasoning, all of which exceeds just following electronic rules. Given the fact that the brain is an *organ with unique neuronal and synaptic connections composed of biomolecular* components *directed by numerous chemicals that produce a great deal of flexibility,* the challenge of trying to duplicate this with just an electrical, digital network proved formidable.

Unlike a computer program, the brain has evolved into various areas representing evolutionary transitions responsible for lesser or more advanced anatomical structures and functions. This includes the reptilian area near the base of the brain that is the source of basic instincts, automatic bodily processes, and behavioral functions; the limbic system or mid-brain that comprises the amygdala, hippocampus, and hypothalamus that together are responsible for memory, emotions, and learning, including much of the hormonal activity of more highly socialized mammals and primates; and the newest, most important convoluted gray matter called the cerebral cortex or cerebrum divided into the frontal, parietal, and occipital lobes that produces such human capacities as language acquisition, learning, reasoning, and creativity.

That Kaku is aware of these differences between computers and human capabilities is indicated in the following statement.

> Given the glaring limitations of computers compared to the human brain, one can appreciate why computers have not been able to accomplish two key tasks that humans perform effortlessly: pattern recognition and common sense. These two problems have defied solution for the past half century. This is the main reason why we do not have robot maids, butlers, and secretaries. (pp. 82–83)

But, as he adds, programmers have been able to overcome these obstacles to some extent. One robot developed at MIT scored higher on object recognition tests than humans, even performing equal to or better than Kaku himself. Another robot named STAIR developed at Stanford University, still relying on the top-down approach, was able to pick out different kinds of fruit, such as an orange, from a mixed assortment that seems simple enough to us, yet very difficult for robots because of the dependence on object recognition. Yet the best result was achieved at New York University where a robot named LAGR was programmed to follow the human bottom-up approach enabling it to identify objects in its path and gradually "learn" to avoid them with increased skill (cf., p. 86).

Furthermore, an MIT robot named KISMET was programmed to respond lifelike to people with given facial expressions that mimicked a variety of emotions (which have now been programmed into dolls), yet "scientists have no illusion that the robot actually feels emotions" (p. 98). While programmers are striving to overcome these differences they still have a long way to go, as Kaku indicates.

> On one hand, I was impressed by the enthusiasm and energy of these researchers. In their hearts, they believe that they are laying the foundation for artificial intelligence, and that their work will one day impact society in ways we can only begin to understand. But from a distance, I could also appreciate how far they have to go. Even cockroaches can identify objects and learn to go around them. We are still at the stage where Mother Nature's lowliest creatures can outsmart our most intelligent robots. (p. 87)

Apparently there are two major approaches to resolving this problem. As indicated previously, Kaku identified two crucial capacities that robots lack that prevent their simulating human behavior: pattern recognition and common sense, both of which require conscious awareness that humans possess and computers and robots entirely lack. One way of solving the problem is to try to endow a computer or robot with consciousness, using a method called "reverse engineering of the human brain." Instead of attempting to "simulate" the function of the brain with an *artificial* intelligence, it involves trying to *reproduce human intelligence* by replicating the neuronal structure of the brain neuron by neuron and then installing them in a robot.

This new method, "called optogenetics, combines optics and genetics to unravel specific neural pathways in animals" (p. 101). Determining by optical means the neural pathways in the human brain presumably would enable optogeneticists not only to *detect* which neural pathways determine specific bodily and mental functions, but also *duplicate* them. At Oxford University Gero Meisenböck and his colleagues

> have been able to identify the neural mechanisms of animals in this way. They can study not only the pathways for the escape reflex in fruit flies but also the reflexes involved in smelling odors. They have studied the pathways governing food-seeking in roundworms. They have studied the neurons involved in decision making in mice. They found that while as few as two neurons were involved in triggering behaviors in fruit flies, almost 300 neurons were activated in mice for decision making. (p. 102)

But the problem is that *identifying* the neuron's function is not the same as *reproducing* it. The intended purpose was to model the entire human brain using two different approaches. The first approach was to "simulate" the vast number of neurons and their interconnections in the brain of a mouse by a supercomputer named Blue Gene constructed by IBM. Computing "at the blinding speed of 500 trillion operations per second . . . Blue Gene was simulating the thinking process of a mouse brain, which has about 2 million neurons (compared to the 100 billion neurons that we have)" (p. 104). But the question is whether simulating is equivalent to reproducing?

This success was rivaled by another group in Livermore, California, who built a more powerful model of Blue Gene called "Dawn." At first in "2006 it was able to simulate 40 percent of a mouse's brain. In 2007, it could simulate 100 percent of a rat's brain (which contains 55 million neurons, much more than the mouse brain" (p. 105). Then progressing very rapidly in 2009 it, "succeeded in simulating 1 percent of the human cerebral cortex . . . containing 1.6 billion neurons with 9 trillion connections" (p. 105).

Although it convinced optogeneticists that *simulating* the human brain was not only possible, it was inevitable, yet again the crucial question is whether "simulating" is equivalent to "reconstructing" or "reproducing," although it seems to me that the distinction has been overlooked and assumed to be the same. Significantly, in addition to meaning "imitating," the term "simulate" has the additional adverse connotations of feigning, pretending, and faking.

The second approach, perhaps to avoid the above problem, is called "reverse engineering of the brain" and it confronted problems of even greater magnitude since it consisted of dissecting the entire system of neurons in the brain into miniscule slices no more than 50 nanometers wide (a nanometer is 1 billionth of a meter) in order to examine each of them under an electron microscope to "reconstruct" their function. Illustrating the enormity of the task, after producing a million slices

> a scanning electron microscope takes a photograph of each, with a speed and resolution approaching a billion pixels per second. The amount of data spewing from the electron microscope is staggering, about 1,000 trillion bytes of data, enough to fill a storage room just for a single fruit fly brain. Processing this data, by tediously *reconstructing* the 3-D wiring of every single neuron of the fly brain, would take about five years. To get a more accurate picture of the fly brain, you then have to slice many more fly brains. (p. 107; italics added)

Although "the human brain has 1 billion neurons more than the fruit fly," it was nevertheless assumed

> that sometime by mid-century, we will have both the computer power to *simulate* the human brain and also crude maps of the brain's neural architecture. But it may take until late in this century before we fully understand human thought or can create a machine that can *duplicate* the function of the human brain. (p. 108; italics added)

Here the distinction between 'simulate' and 'duplicate' seems to be recognized but not considered. Moreover, since we still do not "understand" how the *chemical*-electrical neural processes of the human brain *produce human awareness, perception, memory, emotions, and thought, etc.*, it is questionable whether "constructing" the brain with a wholly *electronic* computer would actually create a "duplicate" of the brain that could function as the original brain.

Among those creating robots there is a consensus that, although in an entirely different way, they can be programmed to "exceed us in intelligence." There is considerable disagreement as to how long this will take, but not that it can be done. According to Kaku a "large part of the problem with these scenarios is that there is no universal consensus as to the meaning of *consciousness*. . . . Nowhere in science have so many devoted so much to create so little" (pp. 110–11). He then offers what he believes are the three capacities essential for being conscious (p. 111):

1. sensing and recognizing the environment
2. self-awareness
3. planning for the future by setting goals and plans, that is, simulating the future and plotting strategy.

I would agree that these are essential aspects, but I do not see what the difficulty has been in attaining a consensus as to the nature of consciousness. When one considers the difference between being awake and being in a *dreamless* sleep or being conscious and then made unconscious by a sedative, blow, or death, we have distinct examples of being conscious and unconscious: in the former cases one is completely aware while in the latter cases one is entirely unaware and unconscious.

There is a difference between the awareness (as minimal as it is) of a fruit fly and a worm and the absence of any awareness in a rose or a rock in that the former involves a sensory content while the latter does not. Of course there are degrees of consciousness, but if one is just sensing, smelling, or feeling one is in a state of minimal awareness. Even dreaming is a kind of pseudo-consciousness in which one is aware that the dream is frightening or pleasant, but that one has no control over it because it is entirely a product of the brain disconnected to one's normal self-awareness and behavioral responses.

That an automaton can be programmed to respond *as if* it had feelings and conscious awareness or to *simulate* either is not sufficient to consider it actually having either. That Deep Blue could defeat Gary Kasparov in a chess match was not an indication that Deep Blue was conscious of the moves it took to defeat the world champion, as Kaku acknowledges. I am not sure whether "self awareness is easier to achieve" as a condition of consciousness than his other two criteria as he claims, but I certainly agree with his assessment of the current state of robotics.

> Today, AI researchers are clueless about how to duplicate all these processes in a robot. Most throw up their hands and say that somehow huge networks of computers will show "emergent phenomena" in the same way that order sometimes spontaneously coalesces from chaos. When asked precisely how these emergent phenomena will create consciousness, most roll their eyes to the heavens. (p. 114)

As previously stated, since it is still a complete mystery as to how the neural discharges in the brain produce consciousness, it is not surprising that AI researchers do not know how electrical circuits, transistors, and computer chips, without any chemical components or regulators, can duplicate or create consciousness. Moreover, while we know that organic evolution produces "emergent phenomena," there is no indication that purely electrical circuits do! When discussing the tremendous progress that has been made in computer science the word "evolution" has been used (or misused), although technical advances have none of the *emergent* features of evolution that is a biological process.

One of the recent breakthroughs in genetics was the discovery of the crucial role the intercellular fluids, composing most of the genome, play in producing the proteins that direct the functions of the individual genes by turning their activities on and off. So is it plausible that even the extraordinary computational advances in artificial intelligence, utilizing only electrical circuits and transistors, can compensate for the lack of these biomolecular, physiochemical, and genetic processes to create robots that will exceed humans in conscious awareness, self-consciousness, intelligence, and creativity? I doubt it.

Roboticists describe the use of prosthesis, inserting electronic devises into the body to restore hearing and vision or replacing amputated limbs with artificial limbs, as heralding a new era in robotics that will integrate electronic devises with biology to create robots with more humanlike capabilities. Kaku cites the example of a twenty-two-year-old young man who had his hand amputated and replaced by "four motors and forty sensors" connected to the nerves in his arm that relay the hand signals to his brain enabling him to "feel" that he had sensations in his hand and control of the movements of his fingers (cf., p. 126).

As remarkable as this is, it must not be overlooked that the *sensations* the young man was feeling in his fingers and his control of them ultimately depended on and occurred in his brain, not in the prosthetic devices that only transmitted the original nerve impulses. People who have had limbs amputated report that they feel sensations in the missing limb, indicating that the sensations are produced by the brain even though felt in the limb—a reminder that the brain is an organ, not a computer.

Yet computer scientists, in a kind of science-fictional world, worry about the day when robots will be produced that are smarter than we are as if computational power were the sole factor in human intelligence. They imagine a world in which the human brain can be duplicated in an electronic system consisting wholly of electrical currents, transistors, and computer chips which, when implanted into a robotic skull, can create a human robot. As reported by Kaku,

> Robot pioneer Hans Moravec . . . explained to me how we might merge with our robot creations by undergoing a brain operation that replaces each neuron of our brain with a transistor inside a robot. The operation starts when we lie beside a robot body without a brain. A robotic surgeon takes every cluster of gray matter in our brain, duplicates it transistor by transistor, connects the neurons to the transistors, and puts the transistors into the empty robot skull. As each cluster of neurons is duplicated in the robot, it is discarded. (p. 130)

Although the tremendous advances in artificial intelligence that disproved the original skepticism about its future achievements should make one cautious about denying its present predictions, I still find this projection, based on dubious assumptions, difficult to understand and accept. Even the Oxford Dictionary defines a nerve impulse as just "a signal transmitted along a nerve fibre consisting of a wave of depolarization," implying it is simply an electronic process. But that overlooks the role of axons and dendrites in connecting the nerve cells via synapses and the neurophysiological and biochemical conditions that facilitate the processes that hardly seems replicable merely by electric circuits and transistors. (For an excellent description of how the various chemicals, hormones, and glands in our brains influence our brain processes and mental states see Joshua Reynolds's excellent account in *20/20 Brain Power.*)[130]

According to Kaku's reported explanation by Hans Moravec, "[a] robotic surgeon takes every cluster of gray matter in our brain, duplicates it transistor by transistor, connects the neurons to the transistors, and puts the transistors into the empty robot skull" while "[we] are fully conscious as this delicate operation takes place." But maintaining that "we are fully conscious during the process" this makes the unlikely assumption that consciousness can be maintained solely by a set of transistors when we do not even know how the normal brain produces it! Also, if the neurons constituting the grey matter in our brains are duplicated by transistors, why must they be reconnected to neurons?

Continuing the description of the operation, Kaku states that following the operation, our brain has been entirely installed into the body of a robot. "Not only do we have a robotic body, we have also the benefits of a robot: immortality in superhuman bodies that are perfect in appearance. This will not be possible in the twenty-first century, but becomes an option in the twenty-second" (pp. 130–31). But this assumes that all life functions can be duplicated entirely by electrical transistors lacking all our neuro*physiological* functions. After all, a robotic body is simply made of metal and silicon completely lacking in the life-giving functions that evolution has endowed humans with!

Moreover, since robots are not capable of reproducing sexually, even if we were "perfect in appearance" what would it matter if we do not have to seek mates to reproduce or even have feelings of sexual attraction and love? Not only would we have lost the happiness of having loving parents and a loving wife, we could never experience the joy and tribulation of having children. And since all who became robots would be immortal, eventually there would be the same group of robotic humans existing forever which, rather than appealing, would seem rather boring since it would be lacking any of the experiences that make human life meaningful, as well as despairing at times. He concludes with this bizarre description:

> In the ultimate scenario, we discard our clumsy bodies entirely and eventually evolve into pure software programs that encode our personalities. We "download" our entire personalities into

> a computer . . . [that] behaves as if you are inside its memory, since it has encoded all your personality quirks inside its circuits. We become immortal, but spend our time *trapped* inside a computer, interacting with other "people" (that is other software programs) in some gigantic cyberspace/virtual reality. (p. 131; [brackets and italics added])

Can one imagine allowing oneself to be reduced to a software program so that one exists inside a computer? What would be the advantage of existing—I can't say "living" because it would not be living in any recognizable way—under such conditions? As a computer program would one have any freedom or control over one's existence? How could we react to other people as mutual "software programs"? Apparently some robots would remain existent to do the downloading, but humans would be "trapped," as Kaku states, in computers. It reads like pure fantasy but roboticists are supposed to be scientists, not fabricators. Being immortal under such conditions would make it even more horrifying. I was relieved when I read Kaku's conclusion: "Some science fiction writers have relished the idea that we will all become detached from our bodies and exist as immoral beings of pure intelligence living inside some computer, contemplating deep thoughts. But who would want to live like that?" (p. 132). No one I am sure; I couldn't agree more!

Turning now to a another recent development: throughout the past it was thought that nonphysical spiritual entities such as the "soul," "life force," Newton's "subtle spirit," Henri Bergson's "élan vitale" were the creative powers in living organisms and thus the force behind evolution culminating in human consciousness, feelings, and thoughts. It wasn't until 1953 when the biophysicist Francis Crick and the geneticist James Watson, aided by the biophysicist Maurice Wilkins and Rosalind Franklin's revealing X-ray crystallographic images of the double helical structure of the DNA, that they were able to eliminate such spiritual explanations. In an article, "A Structure for Deoxyribose Nucleic Acid," they exclaimed that it was the molecular compound, known as DNA, that was the secret of life. Watson, Crick, and Wilkins received the Nobel Prize in 1962 for their achievement (Franklin unfortunately dying four years earlier of ovarian cancer, apparently due to overexposure to the radiation involved in her X-ray crystallography, was ineligible to receive the prize).

It was not until the 1980s, when the theory was finally confirmed, that the earlier psychic or spiritual causes could be definitely rejected. Every molecule of DNA within our cells consists of the twisted ladder-like strands forming a double helix each rung of the helical ladder composed of four nucleic acid bases or nucleotides: adenine (A), thymine (T), cytosine (C), and guanine (G), whose letters convey the genetic code of each individual, along with their ancestral history. Genes are specific sequences in the DNA and RNA that transmit our particular inherited traits and contain the instructions for making proteins. Due to the information processing power of computers, the Human Genome Project, costing 3 billion dollars, succeeded by 2003 in discovering the anatomical blueprint of a human being: the complete sequencing of the genes in each human cell. Since genes are alleged to control about 50% of our neural, physiological, and mental abilities, along with recording our genetic inheritance, they should be considered the "God particles," along with the Higgs Boson.

Again as a result of the extraordinary advances in computational power, it is now possible to encode one's genes on a computer chip or CD. Reading our genetic code enables doctors to identify and forecast the liabilities due to the effects of our genes and, if detected early enough, make it possible to avert various physical and mental disorders. A dramatic discovery showing how genetic research, in addition to disclosing inherited traits, helps explain previously unexplained diseases and abnormalities that can be traced to malfunctioning genes, was announced recently by the National Human Genome Research Institute (NHGRI), a branch of the National Institutes of Health, but involving hundreds of researchers throughout the world.

What they discovered is that the vast amount of intercellular genetic fluids that previously where thought to play no part in the functioning of the genes and thus dismissed as "junk DNA," actually serve a crucial role in directing and regulating the activities of the genes. Composing 98% of the cell substance, instead of being "junk DNA," this fluid substance contains "micro-switches" that convey crucial instructions to the individual genes and their transmissions to other genes. Thus by activating or deactivating a gene's functions they are now considered to be a major factor in producing genetic defects that cause cancer, diabetes, Parkinson's disease, strokes, and heart failure, as well as mental disabilities such as loss of memory, bipolar disorder, schizophrenia, Alzheimer's disease, dementia, and senility.

Having located about four million of these DNA switches, this will enable physicians to learn at a much earlier stage how to diagnose, treat, and eventually prevent the infirmities mentioned above. These findings clearly demonstrate the role of the genome containing the proteins that activate or deactivate the genes and also determine how the chemical modifications of DNA affect gene functions and locate the various operative forms of RNA, another form of nucleic acid similar to DNA, that helps regulate the entire system. Supporting what I wrote earlier about the unlikelihood of the nervous system being replicable by transistors alone, Parkinson's disease is caused by a deficiency of the chemical neurotransmitter dopamine in the brain having the structure $C_8 H_{11}NO_2$.

As a consequence of these enhanced computer investigations a new treatment has been developed, called "tissue engineering," enabling physicians to "grow skin, blood, blood vessels, heart valves, cartilage, bone, noses, and ears in the lab from your own cells" (p. 144). An even more promising discovery was made that has aroused considerable controversy because it involves the destruction of human embryos, a process opposed by the Catholic Church on the grounds that they are living beings. Called "stem cells," they are the earliest cells in the developing human embryo that have not yet differentiated into specialized cells (so they hardly can be considered *human beings*), but have the *potential* to develop into all the various cells of the body.

By injecting these cells into a person with defective organs or who has suffered certain accidents such as spinal cord injury, one will be able to replace the damaged tissue. Presently there are transplants using another person's organs, but given the shortage of replacement organs, the capability of replacing or restoring the defective tissue or organ with engendered stem cells is much more promising. A "pixie dust" has even been created with the power of regrowing tissue.

> This dust is created not from cells but from the extracellular matrix that exists between cells. This matrix is important because it contains the signals that tell the stem cells to grow in a particular fashion. When this pixie dust is applied to a fingertip that has been cut off, it will stimulate not just the fingertip but also the nail, leaving an almost perfect copy of the original finger. Up to one-third of an inch of tissue and nail has been grown in this fashion. (p. 149)

Such are the truly marvellous advances already achieved and others awaiting us!

Another area where there has been considerable medical progress is in reversing aging and increasing our longevity. Kaku reports that medical researchers "have now isolated a number of genes (age-1, age-2, daf-2) that control and regulate the aging process in lower organisms" (p. 168), and since there are counterparts in humans this "has allowed scientists to narrow the search for 'age genes' and look for ways to accelerate the gene repair inside the mitochondria to reverse the effect of aging" (p. 169). He predicts that by 2050,

> it might be possible to slow down the aging process via a variety of therapies, for example, stem cells, the human body shop, and gene therapy to fix aging genes. We could live to be 150 or older. By 2100, it might be possible to reverse the effects of aging by accelerating cell repair mechanisms to live well beyond that. (p. 169)

Some optimists have suggested that when we fully understand the aging process we not only can reverse the process to prolong our lives, but like the robots previously discussed, achieve the religious promise of immortality. But whether or not that would be desirable is another question. In any event it seems unlikely considering that the geological history of the earth indicates that our human species, as all previously advanced species, is doomed to extinction due to drastic climatic changes or the impact of massive meteors or asteroids on the earth unless we can escape to an exoplanet.

Given these amazing neurophysiological, medical, and genetic advances there is now the attempt to explain the problem that has perplexed philosophers since ancient times, usually referred to as the "mind-body problem." Having been so accustomed to directly experiencing thoughts, memories, feelings, emotions, intensions, fears, anger, hate, love, affection, etc., in the past they were simply considered consciousness endowments so different from the body that they were attributed to a soul, vital spirit, or divine endowment. But the recent striking success in *correlating* these mental states with complex neurological structures in our brains has led some researches to consider these underlying neurological processes not just correlated with conscious processes, but the conscious states themselves and thus not needing a separate cause. As Alex Rosenberg, an advocate of this view, states:

> Neuroscience is beginning to answer these questions. We can sketch some of the answer in the work that won Erick Kandel the Noble Prize in Physiology or Medicine. The answer shows how completely wrong consciousness is when it comes to how

> the brain works. Indeed, it shows how wrong consciousness is when it comes to how consciousness works.[131]

Based on his studies of the conditioning formation of the neurons of sea slugs, rats, and humans, Kandel concluded that all our learning and behavioral responses can be explained as due to the evolutionary development producing more neurons with greater molecular complexity and synaptic connections. Continuing Rosenberg's description:

> A little training releases proteins that open up the channels, the synapses, between the neurons, so it is easier for molecules of calcium, potassium, sodium, and chloride to move through their gaps, carrying electrical charges between the neurons. . . . The genes in the nuclei of each cell that control its activities are called somatic genes, in contrast with the germ-line genes in sperm and eggs, which transmit hereditary information. Both kinds of genes contain the same information, since the order of DNA molecules in each of them is the same. Somatic genes are copied from germ-line genes during embryonic development. (p. 181)

Obviously this is a completely reductionistic conception of how we experience the world assisted by how the computer functions as a model for the brain.

> The brain is a computer whose "microprocessors"—its initial assemblies of neural circuits—are hardwired by a developmental process that starts before birth and goes on after it. Long before that process is over, the brain has already started to modify its hardwired operating system and acquired data fed through its sensory apparatus. . . . Beliefs, desires, wants, hopes, and fears are complex information storage states, vast packages of input/output circuits in the brain ready to deliver appropriate and sometimes inappropriate behavior when stimulated. (p. 189)

While my own loss of memory and hearing as I get older is explainable by the deterioration of the areas and functions of the brain correlated with the loss, I find it hard to believe that all the qualitative aspects and impacts of the world and interactions with other people that constitute our experiences and thoughts are nothing more than brain processes. Yet Rosenberg states: "When consciousness convinces you that you, or your mind, or your brain has thoughts [or experiences] about things, it is wrong" (p. 172; brackets added). It seems to me if that were true then we should be experiencing the brain processes themselves, which is quite different from what we normally are aware of.

Moreover, it would deny the referential functions of ordinary language, prose literature, poetry, opera, paintings, applied mathematics, and scientific explanations that are not experienced as neuronal structures of the brain, but mental states referring to or descriptive of the world and enabling us to communicate about it. It is these capacities that enrich our lives and they surely are not *about* the brain but what we are experiencing or thinking. What we see or think when we look at the moon and stars at night are not neuronal discharges but the night sky!

Also the argument that attributing experiences to consciousness is analogous to outmoded explanations in terms of souls, vital spirits, or divine endowments is mistaken because the former were never experienced as such, while the fact of *having conscious experiences and thoughts of our surroundings* can hardly be considered illusory. That our normal experiences are *correlated* with and depend upon the complex chemical-electrical neuronal processes in our brains cannot be denied, but how our brains *produce* our conscious experiences still remains a great mystery.

One of the most dramatic examples of how the brain *produces* extraordinary experiences is that of Joan of Arc whose astonishing religious beliefs and achievements have recently been attributed to a "hyperreligiosity" caused by "temporal lobe epilepsy that can also be induced by what is called "transcranial magnetic simulation" or TMS, along with the epileptic lesions. As described by Kaku in his latest book, *The Future of the Mind*:

> More recently, another theory has emerged about this exceptional woman [Joan of Arc]: perhaps she actually suffered from temporal lobe epilepsy. People who have this condition sometimes experience seizures, but some of them also experience a curious side effect that may shed some light on the structure of human beliefs. These patients suffer from "hyperreligiosity," and can't help thinking that there is a spirit or presence behind everything. Random events are never random, but have some deep religious significance. . . . The neuroscientist Dr. David Engleman says, "Some fraction of history's prophets, martyrs, and leaders appear to have had temporal lobe epilepsy. Consider Joan of Arc, the sixteen-year-old girl who managed to turn the tide of the Hundred Year's War because she believed (and convinced the French soldiers) that she was hearing voices from Saint Michael the archangel, Saint Catherine of Alexandria, Saint Margaret, and Saint Gabriel.[132] (brackets added)

Kaku adds that "Some scientists have gone further and have speculated that there is a "God gene" that predispose the brain to be religious. Since most societies have created a religion of some sort, it seems plausible that our ability to respond to religious feelings might be genetically programmed into our genome" (p. 198). This might explain the universal historical appeal of religions, but not their truth.

Before closing this discussion of how recent science has achieved the major advances in our conception of reality and thus largely replaced religion at least among philosophers and scientists, I think it would be appropriate to continue Kaku's discussion in his previous book of the discovery of the genome and the structure and function of DNA. As previously indicated, DNA is the molecular deoxyribonucleic acid controlling heredity by the genes and, along with the ribonucleic acid RNA, conveys information to proteins directing their essential and ubiquitous functions. These discoveries have been the foremost contributions of genetics to demystifying human origins and revealing our genetic ancestry that disclosed our striking similarity (90% of our genetic makeup correlates with those of mice) to and common origin with other species.

Along with Darwin's evolutionary theory of "natural selection," explaining how our adaptive traits evolved confirmed by the discovery of fossil remains and reinforced by the hereditary evidence encoded in the genome in the nucleus of our cells, they have provided an *entirely naturalistic* explanation of the origin and nature of human beings completely refuting creationism. Moreover, knowing the genetic functions offers a futuristic means of improving human nature.

For the first time in history, due to the decoding of the genome and the structure and function of DNA, RNA, and more recently proteome (the key to explaining the creation of proteins), we now have the means of relieving or remedying the greatest source of human misery. Even more than natural disasters, it is the tyrants, theocrats, terrorists, murderers, rapists, alcoholics, drug addicts, sadists, psychotics, paranoids, and deranged human beings, owing in large part to destructive genes (think of Hitler, Stalin, Hosni Mubarak, Bashar al-Assad, and Putin), who have been and are the major cause of the suffering in the world.

Recall the setting fire to the American Consulate in Benghazi that killed three American diplomats, along with the esteemed Ambassador J. Christopher Stevens who, ironically and tragically, had devoted his life to promoting better relations with Arabic and Muslim nations. Initially explained by US intelligence agents as a reaction to a video defaming Mohammad, after a thorough inquiry a State Department Panel concluded that it was not the video, but the increase in local militia assaults that was the cause that could have been prevented if two State Department Bureaus had responded to requests by officials at the Benghazi Embassy for increased security. Interestingly, following the assaults most shahs and ayatollahs denounced the attacks while thousands of Benghazi residents also demonstrated their opposition.

Of all the various accounts of international conflicts, such as strident nationalism, ethnic conflicts, border and territorial disputes, claims to colonial possessions, economic competition, and ideological or religious disagreements, the one I find must convincing in explaining the uprisings in the Middle East and recent vicious assault on Western embassies by Middle Eastern terrorists is David Deutsch's explanation that they are due to the conflicting contrast between open dynamic and closed static societies in his book *The Beginning of Infinity: Explanations that Transform the World.*[133] He describes dynamic societies as those open to criticisms of the traditional beliefs, institutions, ethnic discriminations, and social and economic discrepancies of their societies often due to changing conditions, acknowledging the right of people to express their convictions according to their individual assessments and dispositions, however unorthodox or erratic, as long as they do not pose a threat to others. It is this tolerance of the critical examination, revision, and rejection of religious doctrines and rituals, along with other repressive cultural and political institutions, while also recognizing the scientific method as the only known reliable method of inquiry that has made many Western countries so progressive. As Carl Sagan in his usual open- minded, astute manner has stated:

> Science is different from many another human enterprise—not, of course, in its practitioners being influenced by the culture they grew up in, nor in sometimes being right and sometimes wrong (which are common to every human activity), but in its passion for framing testable hypotheses, in its search for defini-

> tive experiments that confirm or deny ideas, in the vigor of its substantive debate, and in its willingness to abandon ideas that have been found wanting. If we were not aware of our own limitations, though, if we were not seeking further data, if we were unwilling to perform controlled experiments, if we did not respect the evidence, we would have very little leverage in our quest for the truth.[134]

Despite the opposition of the Catholic Church and some Protestant denominations, consider the progress that has been achieved in Western democratic societies by rejecting such false conceptions, explanations, and institutions as fixed species, geocentrism, *creation ex nihilo,* predestination, rigid hierarchical societies, the divine right of kings, papal or biblical infallibility, and racial or sexual discrimination. In contrast to dynamic societies, static societies (as the West was after the decline of Greek and Roman culture and ascendance of Christianity during the medieval period or dark ages) are closed authoritarian societies opposed to change. The cause of much social unrest and terrorism in the world today, most Islamic states still have theocratic rulers that traditionally repress the civil rights of women and reject practically all democratic reforms due to their acceptance of Muhammad as the prophet of their religion and the Koran as the sacred book believed to be the word of God as dictated to him by the archangel Gabriel.

Thus Muslims generally have rejected scientific inquiry and explanations, such as the Big Bang theory, the theory of evolution, the decoding of the genome with its naturalistic explanation of the origin of human beings, along with the use of contraceptives to avoid unwanted or defective births and overpopulation, and generally are opposed to homosexuals and granting women greater freedom and access to higher education, although recently they were undergoing some liberating changes referred to as the "Arab Spring" that unfortunately has declined.

In dynamic societies, like many in the West today, the separation of state and church has been crucial because it has precluded religious institutions from exerting a controlling influence on society, as has the present opposition of the Catholic Church to same-sex marriages. In democratic or republican forms of government there are documents such as constitutions to define "the rights of man," along with *elected* heads of state such as presidents or prime ministers and parliaments or congresses entrusted to pass laws to enforce these rights and promote social and economic equality with law courts to adjudicate the law. Finally. they insist on access to a free press and uncensored televised news so the public can be well informed and compare their political and social systems to others which has been a major motivation for change, such as the Arab Spring.

Static societies, in contrast, historically have mainly been religious (but include secular dictatorships and fascist regimes) that do not accept the separation of the state and the church or the right of the people to rule. This in turn has justified the repression of free speech and control of the news media to avoid challenges to their authority, along with the enforcement of their laws and restrictions by the fear of imprisonment, torture, and death. Thus the culture of most Islamic countries, excepting the United Arab Emirates, resemble that of the Western medieval period and explains from the Western perspective why they appear to be so backward.

It is this schism between the static societies of the Middle East and the dynamic societies of the West that I believe, as does David Deutsch, to be the major cause of many of the political conflicts in the world today and that has been a deterrent to the democratic liberation and intellectual enlightenment of those ethnic regions. Yet in attributing the conflict to these contrasting differences, it should not be overlooked that religious and political institutions, along with cruel and repressive behaviors, are created by human beings who therefore are ultimately responsible for the resultant conflicts and atrocities and thus any effective remedy must be traced to them.

Consider the imprisonment, torture, and assassination of opposition leaders in Russia, China, Iran, Liberia, and Syria for their advocacy of free speech, an open press, democratic and legal rights. Or recall the murderous assaults of innocent people by deranged individuals in places in the United States such as Aurora, Fort Hood, Tucson, Columbine, and the Sandy Hook Elementary School in Newtown, Connecticut. Recall two bombs that devastated Patriots Day during the Boston Marathon detonated by two immigrant brothers, Tamerlan and Dzhokhar Tsarnaev from the Russian republic of Chechnya, killing four people and causing severe injuries, including numerous amputations to more than 250 others, and the more recent attack in Paris on the newspaper *Charlie Hebdo* and later on a Jewish grocery shop.

Despite all the past changes brought about in the West to enhance its social, economic, democratic, intellectual, and technological developments, no attempt has been made to improve *human nature* because there lacked the knowledge and means to do so, and therefore instilling reasonable, benevolent, humane behavior was entrusted to religious indoctrination, family instructions, ethical teachings, and governmental constraints. But as Kaku graphically describes, with the advances in genetic engineering and the ability to etch everyone's entire DNA on a computer chip or CD, it now will soon be possible to identify the malevolent genes or systems of genes and deactivate, remove, or replace them.

Instead of pouring so much of our financial and scientific resources into developing artificial intelligence and humanoid robots we should devote more of our efforts to improving human nature by detecting and removing the genes causing the depraved behaviors and murderous assaults that pervade society. Just reading the newspaper accounts is extremely depressing. But just as geneticists have discovered the single gene or genes causing such previous fatal afflictions as Parkinson's and Huntington's diseases, diabetes, cystic fibrosis, and hemophilia, and now are on the verge of discovering the genetic cause of various forms of cancer, Alzheimer's disease, schizophrenia and other mental infirmities, they are acquiring the capability to identify the particular genes related to mental derangements that are a primary cause of such abusive behaviors as deadly crimes, vicious rapes, child abuse, sexual trafficking, and tortuous killings. There will be many ethical issues to address and concerns about individual privacy, but this is a social debate well worth having.

This application of genetic engineering to enhancing human nature has been disparaged as attempting to create "designer genes," "recreate eugenics," or rejected as "playing the role of God." But the latter is precisely what scientific knowledge has achieved, as Kaku indicated in showing how scientific attainments have often mirrored the earlier alleged supernatural powers of the gods. In fact, as I write this an analogous example of the contentiousness due to the attempt to remove and replace certain genetically inheritable defects has just been reported in the *Washington Post.*[135] The question of replacing defective genes to eliminate harmful traits from being inherited has already arisen in the recent FDA debate over the possibility of creating three-parent babies called "three-parent IVF." The new procedure would enable mothers who carry mutated or defective DNA in their mitochondria that would transmit such tragic inheritable defects as blindness, epilepsy, schizophrenia, and Down's syndrome to their embryos could have them surgically removed from their extracted egg cells and replaced by refined mitochondria taken from the eggs of a healthy or normal women. Then, after being fertilized in a laboratory, the mother's refined eggs would be replanted in her uterus so that the embryo would not carry the abnormal mitochondrial inheritance.

While this gene replacement procedure would have to be very carefully supervised and regulated, it does not deserve the harsh moral criticism and rejection it has received by some moralists and religionists. Yet such a way of eliminating abominable human traits, analogous to removing sources of horrible diseases and crippling human physical defects, would seem to be another major benefit of genetic discoveries. Consider the tremendous advantages of replacing such inherited genes causing such innate human drives as sadism, pedophilia, hedonism, harmful addictions, avariciousness, vindictiveness, dishonesty, treachery, viciousness, and despotism. Improving human nature and conduct is what I would like to see as the primary achievement of the fourth transition in our revision of reality and way of life. In my view, this would do more to alleviate human suffering and enrich our lives than all other achievements and advances, and it's within our reach.

ENDNOTES

*Galileo Galilei, *The Assayer*, trans. Stillman Drake and reproduced in Stillman Drake and C. D. O'Malley, *The Controversy on the Comets of 1618* (Philadelphia: University of Pennsylvania Press, 1960), p. 3112.

1. Robin Lane Fox, *The Classical World: An Epic History from Homer to Hadrian* (New York: Basic Books, 2006), p. 87; brackets added.

2. Carl Sagan, *The Demon-Haunted World: Science as a Candle in the Dark* (New York: Random House, 1995), pp. 424–26. The subsequent parenthetical citation is to this work.

3. Richard H. Schlagel, *Contextual Realism: A Metaphysical Framework for Modern Science* (New York: Paragon House Publishers, 1986).

4. Aristotle, *Metaphysics*, 1080b16–22; Ross translation.

5. Richard H. Schlagel, *From Myth to Modern Mind: A Study of the Origins and Growth of Scientific Thought*, Vol. I, *Theogony through Ptolemy* (New York: Peter Lang Publishing, Inc., 1995).

6. Plato, *The Republic*, trans. by F. M. Cornford (New York: Oxford University Press, 1945), ch. XXV.

7. Richard McKeon, *The Basic Works of Aristotle* (New York: Random House, 1968), p. 159.

8. Schlagel, *From Myth to Modern Mind*, p. 320.

9. Cf., Charles Singer, *A Short History of Scientific Ideas to 1900* (London: Oxford University Press, 1959), Ch. III. Also see John Boardman, Jasper Griffin, Oswyn Murray, eds., *The Oxford History of the Classical World:* Greece *and the Hellenistic World* (Oxford: Oxford University Press, 1986).

10. T. L. Heath, ed., *The Works of Archimedes with the Methods of Archimedes* (New York: Dover Publications Inc., 19??), pp. 221–22.

11. Henry Osborn Taylor, *The Medieval Mind: A History of the Development of Thought and Emotion in the Middle Ages*, Vol. I, fourth ed. (Cambridge, MA: Harvard University Press, 1962), p. 73.

12. Frederick B. Artz, *The Mind of the Middle Ages* (New York: Alfred A. Knopf, 1962), p. 82.

13. Stephen Greenblatt, *The Swerve: How the World Became Modern* (New York: W. W. Norton & Company, 2011), pp. 92–93.

14. Nicolas Copernicus, *On the Revolutions of the Heavenly Spheres,* trans. by Charles Glenn Wallis (Amherst, NY: Prometheus Books, 1995), p. 6.

15. Letter to David Fabricius, "*Johannes Kepler,*" *Gesammelte Werke,* Vol. XIV, p. 409. Quoted from Arthur Koestler, *The Sleep Walkers* (New York: The Universal Library, 1963), p. 330. The two subsequent parenthetical citations are also to this work.

16. Johannes Kepler, *Astronomia Nova,* "Introduction," *Gesammelte Werke.* Quoted from Arthur Koestler, *The Sleep Walkers,* p. 337.

17. Max Caspar, *Kepler,* trans. and ed. by C. Doris Hellman (New York: Dover Publications, Inc., 1993), p. 286.

18. Johannes Kepler, *Harmonice Mundi,* Book 3, ch. V. R. M. Hutchins, ed., *Great Books of the Western World,* Vol. 16 (Chicago: The University of Chicago Press, 1952), p. 1020.

19. Letter to Herwart von Hohenburg, 1612, 1598, G. W., Vol. XIII, p. 264 *seq.* Quoted from Koestler, *The Sleep Walkers,* p. 340.

20. Richard H. Schlagel, *Forging the Methodology that Enlightened Modern Civilization* (New York: Peter Lang Publishing, Inc. 2011), p. 42.

21. James Gleick, *Chaos: Making a New Science* (New York: Penguin Books, 1988), p. 41.

22. Stillman Drake, *Galileo at Work: His Scientific Biography* (Chicago: The University of Chicago Press, 1978), p. 100.

23. Galileo Galilei, *Sidereus Nuncius* (or The Sidereal Messenger), trans. by Albert Van Helden (Chicago: The University of Chicago Press, 1989), pp. 36–38.

24. Quoted from Stillman Drake, *Galileo at Work,* p. 200. Unless stated otherwise, the subsequent parenthetical citations are to his book.

25. Galileo Galilei, *Discourse on the Comets,* reprinted in *The Controversy on the Comets of 1618,* trans. by Stillman Drake and C. D. O'Malley (Philadelphia: University of Pennsylvania Press), 1960, p. 53.

26. Galileo Galilei, *The Assayer,* reprinted in Drake and O'Malley, *The Controversy on the Comets of 1618,* p. 311. The subsequent parenthetical citation is also to this work.

27. Galileo Galilei, *Dialogue Concerning the Two Chief World Systems—Ptolemaic and Copernican,* trans. by Stillman Drake (Berkeley: University of California Press, 1962), p. 108. Until otherwise indicated, the subsequent parenthetical citations are to this work.

28. Drake, *Galileo at Work,* p. 336. The subsequent parenthetical citations are also to this work until otherwise indicated.

29. Galileo Galilei, *Dialogues Concerning Two New Sciences,* trans. by Henry Crew and Alfonso de Salvio (New York: McGraw-Hill Book Company, Inc., 1963),

p. 147. Unless otherwise indicated, all subsequent parenthetical citations are to this work.

30. Alexander Koyré, *Galileo Studies,* trans. by John Mepham (New Jersey: Humanities Press), 1978. For the quotation and the source for Mersenne, see p. 126, f.n. 177; for Descartes, see p. 107 and source p. 126, f.n. 176.

31. Maurice Clavelin, *The Natural Philosophy of Galileo: Essay on the Origins and Formation of Classical Mechanics,* trans. by A. J. Pomerans (Cambridge, MA: The MIT Press, 1974), p. 383.

32. Richard S. Westfall, *Never at Rest: A Biography of Isaac Newton* (New York: Cambridge University Press, 1983), p. 143. Until otherwise indicated, the subsequent parenthetical citations are to this work.

33. Sir Isaac Newton, *Mathematical Principles of Natural Philosophy,* Vol. I, *The Motion of Bodies,* Motte's trans., revised by Florian Cajori (Berkeley and Los Angeles: University of California Press, 1962), p. xv. Unless or until otherwise indicated, subsequent parenthetical citations are to this work.

34. Sir Issac Newton, *Opticks* or *A Treatise of the Reflections, Refractions, Inflections, & Colours of light* (New York: Dover Publications, Inc., 1952), p. 351.

35. Sir Isaac Newton, *Mathematical Principles of Natural Philosophy and His System of the World,* Vol. II, Motte's trans. revised by Florian Cajori (Berkeley and Los Angeles: University of California Press, 1962), p. 397. Until or unless otherwise indicated, the subsequent parenthetical citations are to this work.

36. Westfall, *Never at Rest,* p. 473. Until otherwise indicated, subsequent parenthetical citations are to this work.

37. I. Bernard Cohen, *Franklin and Newton* (Cambridge, MA: Harvard University Press, 1966), p. 120. Further textual references to this work will be followed by the author's name and page number.

38. Sir Isaac Newton, *The First Book of Opticks,* Book three, Part 1, in *Opticks* or *A Treatise of the Reflections, Refractions, Inflections & Colours of Light.* The subsequent parenthetical citations will be to this work unless otherwise indicated.

39. Schlagel, *From Myth to Modern Mind,* Vol. II, *Copernicus through Quantum Mechanics,* pp. 324–25.

40. William Gilbert, *De Magnete,* trans. by P. Fleury Mottelay (New York: Dover Publications Inc., 1958), p. x; brackets added. Unless or until otherwise indicated, the subsequent parenthetical citations are to this work.

41. Duane Roller and Duane H. D. Roller, *The Development of the Concept of Electric Charge: Electricity from the Greeks to Coulomb,* in James Bryant Conant, General Ed. and Leonard K. Nash, Associate Ed., Vol II, *Harvard Case Histories in Experimental Science* (Cambridge, MA: Harvard University Press, 1970), p. 560. This section includes three drawings of Hauksbee's ingenious apparatuses. Unless or

until otherwise indicated, the immediately following parenthetical citations are to this work.

42. Schlagel, *From Myth to Modern Mind,* Vol. II, *Copernicus through Quantum Mechanics,* p. 330. Until otherwise indicated, all parenthetical citations are to this work.

43. Cohen, *Franklin and Newton,* p. 468.

44. Roller, *The Development of the Concept of Electrical Charge,* pp. 604–605. The subsequent parenthetical citations are also to this work until otherwise indicated.

45. Plato, *The Republic,* Part III, Ch. XXIII, Sec. VI, 509c.

46. Peter Achinstein, *Particles and Waves* (New York: Oxford University Press, 1991), p. 19. The following two parenthetical citations are to this work.

47. Sir Edmund Whittaker, *A History of the Theories of Aether & Electricity,* Vol. I, *The Classical Theories* (New York: Harper & Brothers, 1960), p. 171.

48. Albert Einstein and Leopold Infeld, *The Evolution of Physics* (New York: Simon and Schuster, 1951), pp. 155–56.

49. Joseph Priestley, *Experiments on Air,* 1790, Vol. I, p. 248. Quoted from J. R. Partington, *A Short History of Chemistry,* third ed. Revised and enlarged (New York: Harper and Brothers 1960), p. 137. The following three parenthetical citations are to this work.

50. James Bryant Conant, *The Overthrow of the Phlogiston Theory,* in James Bryant Conant, General Ed. and Leonard K. Nash, Associate Ed., *Harvard Case Histories in Experimental Science,* Vol. I (Cambridge, MA: Harvard University Press, 1948), pp. 69–70.

51. Elizabeth C. Patterson, *John Dalton and the Atom Theory: The Biography of a Natural Philosopher* (New York: Anchor Books, 1970), p. 21. This summary of Dalton's early life and the textual references are largely based on her excellent study. Subsequent parenthetical references are to this work.

52. Leonard K. Nash, *The Atomic-Molecular Theory,* in James Bryant Conant, General Ed. and Leonard K. Nash, Associate Ed., *Harvard Case Histories in Experimental Science,* Vol. 1, p. 222. The immediately following parenthetical reference is also to this work.

53. Frank Greenaway, *John Dalton and the Atom* (Ithaca: Cornell University. Press, 1966), p. 133. The following quotation is also to this work.

54. This discussion of J. J. Berzelius is based on Bill Bryson, *A Short History of Nearly Everything* (New York: Broadway Books, 2003), pp. 105–106.

55. Nash, *The Atomic-Molecular Theory,* p. 248. All of the subsequent extensive numerical citations are to the work.

56. Dmitri Ivanovich Mendeleev, Faraday Lecture, 1889, *The Principles of*

Chemistry, Vol II, in William C. Dampier and Margaret Dampier (eds.), *Readings in the Literature of Science* (New York: Harper Torchbooks, 1959), p. 115.

57. Michio Kaku, *The Future of the Mind* (New York: Doubleday, 2014), p. 133.

58. Albert Einstein and Leopold Infeld, *The Evolution of Physics*, p. 129.

59. Peter Achinstein, *Particles and Waves*, pp. 17–19.

60. Isaac Newton, *Principia Mathematica*, Vol. II, p. 547.

61. Singer, *A Short History of Scientific Idea to 1900*, p. 360. The subsequent parenthetical citation is also to this work.

62. G. Kirchhoff and Robert Bunsen, *Annals of Physics and Chemistry* 110 (1860): 160; trans. in *Philosophical Magazine*, 20 (1860): 89. Quoted from Abraham Pais, *Inward Bound* (Oxford University Press, 1988), p. 168. Until otherwise indicated, the subsequent parenthetical citations are to this work.

63. J. J. Thomson, "Cathode Rays," *Philosophical Magazine* 44 (August 7, 1897), p. 311.

64. Richard R. Schlagel, "The Waning of the Light: The Eclipse of Philosophy," *Review of Metaphysics* 57 (September 2003): 105.

65. Emilio Segrè, *From X-Rays to Quarks: Modern Physicists and Their Discoveries* (San Francisco: W. H. Freeman and Company, 1980), p. 47. The following two parenthetical citations are to this work.

66. Pais, *Inward Bound*, p. 112. The following quotation is also to this work.

67. Max Planck, *Scientific Autobiography and Other Papers*, trans. by G. Gaynor (London: William and Norgate, 1950), p. 7. Quoted from Abraham Pais, *Niels Bohr's Times: In Physics, Philosophy, and Polity* (Oxford: Clarendon Press, 1991), p. 83.

68. Armin Hermain, *The Genesis of Quantum Theory* (Cambridge, MA: MIT Press, 1971), p. 23. Quoted from Segrè, *From X-rays to Quarks*, p. 76. The following three parenthetical citations are also to this work.

69. Planck, *Scientific Autobiography and Other Papers*, p. 7. Quoted from Abraham Pais, *Niels Bohr's Times*, p. 86.

70. Albert Einstein, *Albert Einstein: Philosopher-Scientist*, "Autobiographical Notes," ed. by Paul Arthur Schilpp (Evanston, IL: The Library of Living Philosophers, Inc., 1949), p. 45.

71. Jean Perrin, *Brownian Movement and Molecular Reality*, trans. by F. Soddy (London: Taylor and Frances, 1910), concluding paragraph. Quoted from Abraham Pais, *"Subtle is the Lord . . .": The Science and the Life of Albert Einstein* (Oxford: Oxford University Press, 1982), p. 95.

72. For a schematic representation of the experiment see Leo Sartori, *Understanding Relativity* (Berkeley and Los Angles: University of California Press, 1996), p. 30.

73. Albert A. Michelson, *American Journal of Science* 22 (1881): 120. Quoted

from Abraham Pais, "*Subtle is the Lord . . . ,*" p. 112. The subsequent parenthetical citation is to this work.

74. Albert Einstein, *Relativity: The Special and General Theory* (New York: Crown Publishers, Inc., 1961), pp. 32–34.

75. G. J. Whitrow, *The Structure and Evolution of the Universe* (New York: Harper Torchbooks, 1959, p. 85.

76. Milic Capek, *The Philosophic Impact of Contemporary Physics* (New York: D. Van Nostrand Company, Inc., 1961), p. 201.

77. Pais, "*Subtle is the Lord . . . ,*" p. 178.

78. Albert Einstein, *James Clerk Maxwell* (Cambridge: Cambridge University Press, 1931), p. 66. Quoted from Abraham Pais, *Inward Bound,* p. 244.

79. Pais, *Inward Bound,* p. 189.

80. Jean-Baptiste Perrin, *Review Scientifique* 15 (1901): 447. Quoted from Pais, *Inward Bound,* p. 183.

81. Segrè, *From X-Rays to Quarks,* p. 136.

82. Yuval Ne'eman and Yoram Kirsh, *The Particle Hunters* (Cambridge: Cambridge University Press, 1996), p. 14. The subsequent parenthetical quotation is also to this work.

83. *Philosophical Magazine* 37 (1919): 581. Quoted from Segrè, *From X-Rays to Quarks,* p. 110.

84. Niels Bohr, *Philosophical Magazine* 25 (1913): 10. Quoted from Pais, *Niels Bohr's Times,* p. 128.

85. Pais, *Inward Bound,* p. 198. The subsequent parenthetical citation is to this work.

86. Robert P. Crease and Charles C. Mann, *The Second Creation: Makers of the Revolution in Twentieth-Century Physics* (New York: Macmillan Publishing Company, 1986), p. 27.

87. Pais, *Inward Bound,* p. 199.

88. Pais, *Niels Bohr's Times,* p. 152.

89. Einstein, *Albert Einstein: Philosopher-Scientist,* pp. 45–47.

90. Abraham Pais, *Niels Bohr's Time,* p. 152.

91. Ne'eman and Kirsh, *The Particle Hunters,* p. 37. The subsequent parenthetical citation is also to this work.

92. Werner Heisenberg, *Physics and Beyond: Encounters and Conversations,* trans., from the German by Arnold J. Pomerans (New York: Harper and Row Publishers Inc., 1972), p. 38. Until otherwise indicated, the subsequent parenthetical citations are to this work.

93. Crease and Mann, *The Second Creation,* p. 50. The subsequent parenthetical citation is to this work.

94. Ne'eman and Kirsh, *The Particle Hunters*, p. 44. The following two parenthetical quotations are to this work.

95. Crease and Mann, *The Second Creation*, pp. 52–53. I have rearranged the brackets to make the quotation more grammatical.

96. J. C. Polkinghorne, *The Quantum World* (Princeton, NJ: Princeton University Press, 1985), p. 30. For those who are interested, in addition to giving the formula, he describes the meaning of the different mathematical symbols more fully.

97. Pais, *Niels Bohr's Times*, p. 285. In footnotes Pais gives the sources for this quotation.

98. Crease and Mann, *The Second Creation*, p. 55. The immediately following two parenthetical quotations are to this work.

99. Pais, *Niels Bohr's Times*, p. 286. The subsequent parenthetical quotation is to this work.

100. Max Born, *My Life and My Views* (New York: Charles Scribners and Sons, 1968), p. 55.

101. Segrè, *From X-Rays to Quarks*, p. 165.

102. Pierre-Simon Laplace, *Inroduction à la théorie analytique des probabilities* (Paris: *Oeuvres Complètes*, 1886), p. VI. Quoted from Capek, *The Philosophic Impact of Contemporary Physics*, p. 122.

103. Quotation from Abraham Pais, *Niels Bohr's Times*, p. 309.

104. Werner Heisenberg, *Physics and Philosophy: The Revolution in Modern Science* (New York: Harper & Brothers Publishers, 1958), p. 42.

105. Albert Einstein, Boris Podolsky, and Nathan Rosen, "Can Quantum Mechanical Descriptions of Physical Reality Be Considered Complete?" *Physical Review* 47 (1935): 777–80.

106. Niels Bohr, "Can Quantum-Mechanical Descriptions of Physical Reality Be Considered Complete?" *Physical Review* 48 (1935): 696–702.

107. Niels Bohr, *The Philosophical Writing of Niels Bohr*, Vol. III, *Essays 1958–62 on Atomic Physics and Human Knowledge* (Woodbridge: Ox Bow Press, 1987), p. 4.

108. Ne'eman and Kirsh, *The Particle Hunters*, p. 53. My simplified brief discussion of spin and the following description of the "Pauli Exclusion Principle" are based on this work as are the following three parenthetical citations, though I am responsible for any oversimplifications or misinterpretations. Describing these more recent quantum mechanical developments is much more difficult because of their greater complexity and the fact that their dependence on the mathematical formulation makes it difficult to render it in ordinary language. The following three mathematical quotations are to this work.

109. Crease and Mann, *The Second Creation*, p. 82. The subsequent parenthetical citation is to this work.

110. Pais, *Inward Bound*, p. 290.

111. Segrè, *From X-Rays to Quarks*, p. 171.

112. Crease and Mann, *The Second Creation*, p. 83. The subsequent parenthetical citation is also to this work.

113. Chris Quigg, "Elementary Particles and Forces," *Scientific American* (April 1985): 83.

114. Ne'eman and Kirsh, *The Particle Hunters*, p. 59. The following five parenthetical citations are to this work.

115. Cf. George Johnson, *Strange Beauty* (New York: Vintage Books, 2000), pp. 267–96 for an excellent summary of this development. The immediately following parenthetical references are to this work unless or until otherwise indicated.

116. Crease and Mann, "How the Universe Works," *Atlantic Monthly* (August 1984: 91.

117. Steven Weinberg, *Dreams of a Final Theory: The Scientist's Search for the Ultimate Laws of Nature* (New York: Vintage Books, 1994), p. 237.

118. Harald Fritzsch, *Quarks: The Stuff of Matter* (New York: Basic Books, Inc., 1983), p. 10.

119. Crease and Mann, *The Second Creation*, p. 410.

120. For a review of *The Grand Design* see Steven Weinberg's article, "The Universe We Still Don't Know," *New York Review of Books*, February 10, 2011.

121. Bill Bryson, *A Short History of Nearly Everything* (New York: Broadway Books, 2003), pp. 168–69. The following three quotations are also to this work.

122. Albert Einstein, *Ideas and Opinions* (New York: Bonanza Books, 1964), p. 266.

123. Albert Einstein, quoted from Manjit Kumar, *Quantum* (New York: W. W. Norton & Company, 2010), p. 320.

124. Brian Vastag and Joel Achenbach, "Scientists Laud Particle Discovery," *Washington Post*, July 8, 2012, A1. The following three parenthetical citations as A2 are to this article.

125. John Horgan, *The End of Science: Facing the Limits of Knowledge in the Twilight of the Scientific Age* (New York: Broadway Books, 1997), p. 231.

126. Brian Vastag, "WIMPs: Hard to See but Vital to the Cosmos," *Washington Post*, December, 4, 2012, pp. E1 and E5. The following parenthetical citations are also from E-5.

127. Kaku Michio, *Parallel Worlds* (New York: Doubleday, 2005), p. 191. Until otherwise indicated, all the subsequent parenthetical citations are also to this work.

128. Alex Rosenberg, *The Atheist's Guide to Reality: Enjoying Life Without Illusions* (New York: W. W Norton & Company, Inc. 2012), pp. 25–26.

129. Michio Kaku, *Physics of the Future: How Science Will Shape Human Destiny and Our Daily Lives by the Year 2100* (New York: Anchor Books, 2011). Except for two additional citations, all of the subsequent parenthetical quotations are to this work.

130. Joshua Reynolds with Robert Heller, MD, *20/20 Brain Power* (Laguna Beach, CA: 20/20 Brain Power Partners, LLC, 2005), chaps. 8 and 10.

131. Rosenberg, *The Atheist's Guide to Reality*, p. 180. The immediately following parenthetical references are also to this work.

132. Michio Kaku, *The Future of the Mind*, p. 196.

133. David Deutsch, *The Beginning of Infinity: Explanations that Transform the World* (New York: Viking Penguin, 2011), pp. 379–88.

134. Sagan, *The Demon-Haunted World*, p. 263.

135. Ariana Eunjung Cha and Sandhya Somashekhar, "FDA Panel Debates Idea of Three-parental Babies," *Washington Post*, February 26, 2014, pp. A1, A4.

INDEX